Arctic Opportunities and Challenges

Edmund Li Sheng

Arctic Opportunities and Challenges

China, Russia and the US Cooperation and Competition

Edmund Li Sheng
University of Macau
Macau, China

ISBN 978-981-19-1245-0   ISBN 978-981-19-1246-7   (eBook)
https://doi.org/10.1007/978-981-19-1246-7

This Palgrave Macmillan imprint is published by the registered company Springer Nature
Singapore Pte Ltd.
The registered company address is: 152 Beach Road, #21-01/04 Gateway East, Singapore
189721, Singapore

CONTENTS

CHAPTER 1

Introduction: The Game in the Arctic—A Historical Review

Abstract This work mainly reviews, discusses, and looks toward the competition and cooperation between the three global powers, China, the United States and Russia, in the Arctic region, from the perspective of triangle relations theory in international politics.

Keywords Great Triangle Theory · Sino-US Relations · Sino-Russia Relations

For more than a century, the Arctic has been an arena for the game of the great powers because of its abundant natural resources. Originally, there were eight direct players in this game based on their geographic proximity, and these are known as "the Eight Arctic Countries": Russia, Canada, the United States, Denmark capital (Greenland), Norway, Finland, Iceland, and Sweden. In the Cold War era, the Arctic was a frontline in the bipolar confrontation between the United States and the Soviet Union, and this bipolar system is still in place in the post-Cold War era. As neighbors eyeing each other across the Arctic Ocean, the Arctic is crucial for both the United States and Russia. The significance of the Arctic did not decrease even after the collapse of the Soviet Union, and the end of the Cold War did not mark the end of Arctic competition.

Competition in the Arctic has however become increasingly complicated in recent decades. With the trend of global warming, the unprecedented speed with which the polar ice caps are melting is changing the landscape of the Arctic, both literally and in geopolitical terms. The melting polar ice opens the door wider for human access to the region around the North Pole. It not only lowers the cost of exploiting polar resources but also opens more maritime routes connecting Western Europe, East Asia, and North America, thus increasing the convenience of international trade for three of the most significant regions in the global economy. In these circumstances, it can be predicted that the Arctic will again be an important arena for the great powers and for some of the world's key economies, which will deeply transform the regional and global geopolitical landscape just as physical changes mark the Arctic landscape.

In the context of global warming, the affairs of the Arctic are no longer exclusively the concerns of the member states of the Arctic Council. Players from outside of the region of the traditional eight Arctic states are keen to participate in the geopolitical game, including the European Union, China, Japan, and South Korea. With its overlapping membership and significant security concerns, the European Union takes part directly and indirectly in the Arctic game, whereas the geographic distance of the three abovementioned East Asian countries leave them as outsiders.

By dint of the successful export-oriented strategy, China have enjoyed impressive economic growth over the past few decades at an annual growth rate over 6% and has overtaken Japan to become the world's second-largest economy in 2010.[1] As the most important trading nation, China attaches great importance to the geopolitical situation in the Arctic. In recent decades, not being an Arctic country has not prevented China from expressing its lasting interest in Arctic affairs. In addition to showing its geopolitical ambition, China's proactive movements in the region are driven by security issues. It is usually believed that by developing routes in the North, China aims to reduce the reliance on traditional southern maritime routes, especially through the Strait of Malacca and the South China Sea, where the geopolitical risks are relatively high.

---

[1] Sheng, L. (2014b). Economic structure, cost outsourcing and global imbalances. *Journal of Australian Political Economy, 74*, 81–94.

Over the last few years, China has made a series of steps toward enhancing its presence in the Arctic. In 2013, China became an official observer state of the Arctic Council and attempted to increase its cooperation with member states. In 2017, the National Development and Reform Commission (NDRC) and the State Oceanic Administration (SOA) jointly issued the Vision for Maritime Cooperation under the Belt and Road Initiative, proclaiming the goal to build a peaceful and prosperous 21st Century Maritime Silk Road. In the same year, China and Russia started to cooperate on building "the Polar Silk Road," which announced the increasing presence of China in the Arctic region and triggered concern on the part of the United States and its European allies. In the following years, China expressed an even stronger interest in the region in its official Arctic white paper, which positioned the country as a "near-Arctic state" and "an important shareholder in Arctic affairs."

For a comprehensive understanding of China's Arctic policy, it is important to take a panoramic view of China's foreign policy in recent years, especially the Belt and Road Initiative (BRI). China has long kept its capital account largely closed to avoid financial risks and economic volatility.[2,3] The implementation of the BRI is aimed at further enhancing China's economic openness and attracting more foreign direct investment (FDI). Indeed, the Polar Silk Road is an extension of the BRI into the region around the North Pole, through which is possible to uncover a new geopolitical structure operating in the Arctic region: With the increasingly strong presence of China in the Arctic, the bipolar competition between Russia and the United States is generally shifting into a triangular game. Just as it did in the 1970s, China is again playing the role of game changer in the bipolar competition between Russia and the United States by adding a third player to the game.

This work mainly reviews, discusses, and looks toward the future of the competition and cooperation in the Arctic region between the three global powers—China, the United States, and Russia—from the perspective of the theory of triangular relations in international politics. In modern international relations, the term "strategic triangle" was proposed to refer to Sino–American–Soviet relations during the 1970s.

[2] Sheng, L. (2012). Dealing with financial risks of international capital flows: A theoretical framework. *Cambridge Review of International Affairs, 25*(3), 463–474.

[3] Sheng, L. (2014a). Capital controls and international development: A theoretical reconsideration. *Global Policy, 5*(1), 114–120.

This triangle still operates in the relations between China, the United States, and Russia and shapes the international arena. These three key powers may be competitors in one field and partners in another, which is why their mutual influence is of such importance. China's Arctic ambitions inevitably trigger concerns among the Arctic states, considering their strong resistance against the participation of non-Arctic countries in Arctic affairs, but its presence in the region could be more profound in the future with the formation of a new strategic triangle in the Arctic. Against the background of Sino–American competition, Artic affairs are a high-level test of China's international strength, with China positioned as the extra-regional actor and the United States as the direct participant. Meanwhile, China's entry into the region could leave a door open for other East Asia countries to become involved in Arctic trade routes and thus change the landscape of international shipping, trade, and even security.

The theoretical framework of this book encompasses the general theory of geopolitics and the specific theory of a strategic triangle formed by China, Russia, and the United States. We argue that the game played over the Arctic by China, Russia, and the United States epitomizes a great strategic triangle between these three major powers in today's world. To establish this point of view, we first present a literature review on the origins of this strategic triangle, dating back to the 1970s. To fill a literature gap on Sino–American–Soviet Union/Russian relations and interactions in the Arctic region, we then compare the earlier triangular relations between China, the Soviet Union, and the United States with the current relations between China, Russia, and the United States. We also adopt a historical approach to summarize the interactions of the great powers in the region around the North Pole from the Cold War to the post-Cold War eras. In particular, we analyze these interactions in the context of China's geopolitical strategy, including the BRI and Polar Silk Road, from the wider perspective of economics, international trade, and military affairs.

In the first section of the book, we provide a panoramic view of the Arctic by illustrating its geographic significance, legal status, and regional disputes. In the first chapter, we depict the evolution of Arctic competition over the past half-century and introduce the Arctic policies of the key players: the United States, Russia, the European Union, the United Kingdom, Japan, and South Korea. We predict that, with so many players involved in this game, competition over the Arctic will be constantly fierce and will cast a shadow over attempts at multilateral cooperation and development in the region. The Arctic ambitions of extra-regional players like China will face a series of challenges. Limited by geographic distance, China can only begin to carry out its Arctic ambitions through cooperation with Arctic countries, despite the loss of initiative that this entails. In the meantime, the exclusiveness of the Arctic Council and the regional countries is a major hindrance for China and other extra-regional players to become involved in Arctic affairs. For China in particular, the circumstances are not advantageous as there are other extra-regional players, such as Japan and South Korea that are better situated because of their international identity and relatively advanced technology. We follow this line of thinking in anticipating China's future Arctic policy, whether its participation in Arctic affairs is likely to be smooth, and what its future directions might be from different international perspectives.

REFERENCES

Sheng, L. (2012). Dealing with financial risks of international capital flows: A theoretical framework. *Cambridge Review of International Affairs, 25*(3), 463–474.

Sheng, L. (2014a). Capital controls and international development: A theoretical reconsideration. *Global Policy, 5*(1), 114–120.

Sheng, L. (2014b). Economic structure, cost outsourcing and global imbalances. *Journal of Australian Political Economy, 74*, 81–94.

CHAPTER 2

A Panorama of the Arctic: Geopolitics and International Law

Abstract As the Arctic has exhibited the effects of rapid climate change over the past few decades, it has become a new focus of competition between states. This increased competition has unfolded as various countries have sought to maximize their own economic and political benefits in the region. This chapter provides a panorama of the rapidly changing Arctic by illustrating its geographic significance and legal status and outlining the disputes over the region.

Keywords Arctic · Geopolitics · International relations · Territorial disputes · Multilateralism

What the Aegean Sea was to classical antiquity, what the Mediterranean was to the Roman world, what the Atlantic Ocean was to the expanding Europe of Renaissance days, the Arctic Ocean is becoming to the world of aircraft and atomic power.
—Hugh Llewellyn Keenleyside, 1942

E. L. Sheng, *Arctic Opportunities and Challenges*,
https://doi.org/10.1007/978-981-19-1246-7_2

2.1   RAPID CHANGES IN THE GEOPOLITICS OF THE ARCTIC: FROM OBSERVATIONS TO THEORY

Despite its location on the geographic periphery of the globe, the Arctic occupies a central position in international geopolitics. In many works, the Arctic Ocean is metaphorically compared with the Mediterranean, which reveals the geopolitical complexity of this region. Indeed, the two seas have physical and historical similarities: according to Tamnes and Holtsmark, "both are relatively shallow, with narrow and defensible choke points," and "both are, or have been tactically and strategically important for the projection of naval, air, and land-based military power from one continent to another."[1]

Tamnes and Holtsmark also teased out the geopolitical history of the Arctic. During the inter-war period, the British Navy set its sights on the Arctic and made efforts to explore a navigable route through the Northwest Passage, thereby helping to map the region.[2] During the Cold War, with the development of strategic bombers, intercontinental missiles, and submarines, the Arctic was highly militarized and became a veritable strategic theater for the superpowers to display their capabilities in the interest of balance and deterrence. In the post-Cold War era, the geopolitical status of the Arctic experienced a transition from the strategic frontline to "a regional subsystem encompassing a variety of states with varying cross-border relationships that range from strategic competitions to collaboration, and often display a seemingly paradoxical blend of competition and collaboration."[3] In the recent decade, the Arctic has once again attracted the world's attention with the melting of its ice due to intensifying global warming. On August 21, 2007, the Northwest Passage became open to ships without the need of an icebreaker, and the Northeast Passage also opened one year later.[4]

---

[1] Tamnes, R., & Holtsmark, S. G. (2014). The geopolitics of the Arctic in historical perspective. In R. Tamnes & K. Offerdal (Eds.), *Geopolitics and security in the Arctic* (pp. 26–62). Routledge.

[2] Ibid.

[3] Ibid.

[4] NSIDC. (2008). *Sea ice decline accelerates, Amundsen's Northwest Passage opens.*

However, as Zellen[5] argued, the "Big Thaw" in the Arctic is a double-edged sword. On the one hand, "it will open the region to a 'Cold Rush' of economic and military exploitation, as long-sought sea lanes and the vast hydrocarbon riches of the Arctic seafloor will be exposed by the shrinking of the ice cap for the first time in human history." On the other hand, it will not only increase the possibility for environmental disaster affecting many Arctic biota but also pose a series of challenges to the subsistence economy and traditional cultures of the indigenous peoples of the Far North.

If we turn to observe the Arctic from the perspective of geopolitics, we find that as a specific geographical area it has always been regarded as an independent or semi-independent system,[6] the interaction of which with the international community is not considered historically significant. With the rapid changes in the global climate, however, this once isolated and forgotten region has again become a research focus for scholars of international relations and political economy.

In highlighting the specificities of the Arctic region, Heather described the Arctic as fundamentally a regional security complex built around the interdependence of environmental and marine issues. These interdependent issues have an impact on what needs to be addressed in relation to the Arctic and the regional public goods that the Arctic can provide. In the decades following the Cold War, the Arctic shifted rapidly from a marginal military zone to an international political focal point. The image of the Arctic in the international arena also changed dramatically, from a passive object to an active subject.[7] As competition for natural resources has become one of the core features of the Arctic, jurisdictional conflicts among major powers, especially China, the United States, and Russia, have become fierce.[8] At the same time, Arctic states have begun to face

---

[5] Zellen, B. S. (2009). *Arctic doom, Arctic boom: The geopolitics of climate change in the Arctic.* ABC-CLIO.

[6] Wegge, N. (2011). The political order in the Arctic: Power structures, regimes and influence. *Polar Record, 47*(2), 165–176.

[7] Exner-Pirot, H. (2013). What is the Arctic a case of? The Arctic as a regional environmental security complex and the implications for policy. *The Polar Journal, 3*(1), 120–135.

[8] Sheng, L. (2021) *How Covid-19 reshapes new world order: Political economy Perspective.* Springer.

more maritime security problems, and the region has begun to face non-traditional security issues, such as environmental degradation and resource overexploitation. In this context, the competition among major countries in the Arctic is essentially a driver of and challenge to Arctic governance mechanisms.

The competition among the major states in the Arctic today, especially between the three major powers of China, the United States, and Russia, actually represents a conflict between two different styles of mechanism for Arctic governance: namely statism and globalism. The Arctic states, with their dominance of the Arctic Council, remain the central force in Arctic governance, with extraterritorial states having minimal substantive power to engage in Arctic affairs. State-centrism, with a focus on territory and jurisdiction, therefore remains dominant in Arctic governance. The strategy documents of all Arctic states emphasize the importance of state sovereignty and the means to preserve it.[9] In practice, the Arctic states seek to monopolize the structure of Arctic governance in the Arctic Council, through both collective collaboration on Arctic affairs and active bilateral and multilateral cooperation among these countries.

According to the principle of state-centrism, the Arctic states are the only legitimate participants in the governance of Arctic affairs. However, in an era of increasing interdependence associated with globalization, the boundaries of the Arctic region are blurring, both geographically and in terms of the global scope of its influence. Accordingly, the governance mechanism of the Arctic is flattening, with a trend toward the globalization of Arctic affairs. Economic structures and external policies of countries are deeply intertwined and influence each other in the globalization context.[10,11] As a part of this trend, major powers without immediate geographic proximity to the Arctic are expressing ambitions to participate in Arctic governance and are exerting a certain influence on the decision-making mechanisms. A vivid example of the practice of globalism in the Arctic is China's proposed "Polar Silk Road." The high rates of economic growth in the country resulting from the expansion of

[9] Heininen, L. (2012). State of the Arctic strategies and policies–A summary. *Arctic Yearbook 2012*, 2–47.

[10] Sheng, L. (2012). Dealing with financial risks of international capital flows: A theoretical framework. *Cambridge Review of International Affairs*, 25(3), 463–474.

[11] Sheng, L. (2014a). Capital controls and international development: A theoretical reconsideration. *Global Policy*, 5(1), 114–120.

manufacturing exports have provided a foundation for China to export its influence.[12,13] It is a complement to China's Belt and Road Initiative which the US labels as a nonmarket economic model China try to extend to other countries.[14]

Corneliu Bjola's proposal for Arctic plurilateral cooperation seems a good fit for China's Arctic strategy.[15] Unlike regional arrangements, which limit the participation of governance to intraterritorial actors, plurilateralism breaks down geographical constraints by allowing extraterritorial actors to participate in discussions on issues of common interest. Furthermore, in contrast to multilateralism, in which sovereign states are the core actors, plurilateralism is issue based and emphasizes international cooperation among multiple types of actors, including governments, interest groups, and non-governmental organizations (NGOs). Plurilateralist initiatives provide an avenue for Asian countries such as China to participate in Arctic-specific issues alongside the Arctic states and also provide a framework for further discussion of the strategic triangle in operation between China, the United States, and Russia in the Arctic. Against this background, the Arctic is becoming another international theater in which the problems of international competition and conflict will be posed to the regimes of international law and multilateralism in world politics.

2.2 The Legal Status of the Arctic

Over the past century, the polar regions have attracted ever-increasing attention across the globe as the result of geopolitical factors, technological advances, and the intensification of global warming. As a result, a series of international regimes have been built up for cooperation on their management.

[12] Sheng, L. (2014b). Economic structure, cost outsourcing and global imbalances. *Journal of Australian Political Economy, 74*, 81–95.

[13] Sheng, L. (2015). Theorizing income inequality in the face of financial globalization. *The Social Science Journal, 52*(3), 415–424.

[14] Sheng, L., & do Nascimento, D. F. (2021). *Love and trade war: China and the US in historical context*. Springer.

[15] Bjola, C. (2013). Keeping the Arctic "cold": The rise of plurilateral diplomacy? *Global Policy, 4*(4), 347–358.

In 1959, the Antarctic Treaty was signed in Washington by the 12 countries that had been active in scientific pursuits in and around Antarctica. The treaty came into force in 1961 and laid down a solid foundation of principles for international activity in Antarctica, including peaceful purposes, freedom of scientific investigation, international scientific cooperation, territorial sovereignty, nuclear activity, geographical coverage, inspections, and jurisdiction. The treaty also defined its legal status and what was meant by the peaceful purposes of the use of Antarctica:

> Antarctica shall be used for peaceful purposes only. There shall be prohibited, inter alia, any measures of a military nature, such as the establishment of military bases and fortifications, the carrying out of military maneuvers, as well as the testing of any type of weapons.
>
> Nothing contained in the present Treaty shall be interpreted as: a renunciation by any Contracting Party of previously asserted rights of or claims to territorial sovereignty in Antarctica; a renunciation or diminution by any Contracting Party of any basis of claim to territorial sovereignty in Antarctica which it may have whether as a result of its activities or those of its nationals in Antarctica, or otherwise; prejudicing the position of any Contracting Party as regards its recognition or non-recognition of any other States's rights of or claim or basis of claim to territorial sovereignty in Antarctica.[16]

Compared with the continent of Antarctica, the situation in the Arctic is more complicated in terms of its fragmented geography and legal regimes. The Arctic region consists of various parts, including the North Pole, the Arctic Ocean, the territory of the eight countries of the Arctic Circle, and "an imaginary line that marks the latitude above which the sun does not set on the day of the summer solstice and does not rise on the day of the winter solstice."[17] More specifically, "the Arctic Region includes the five states surrounding the Arctic Ocean—Canada, Denmark (Greenland), Norway, Russia, and the United States—and the straits used for international navigation to and from the Arctic Ocean—Bering Strait, Northwest Passage, Northeast Passage/Northern

---

[16] The Antarctic Treaty. (1959).

[17] Clote, P. (2008). Implications of global warming on state sovereignty and Arctic resources under the United Nations Convention on the Law of the Sea: How the Arctic is no longer communis omnium naturali jure. *Richmond Journal of Global Law and Business, 8*, 195.

Sea Route, and the Nares, Davis, Fram and Denmark Straits. Another three states have land territory north of the Arctic Circle (66°33'39"N)— Iceland, northern Sweden, and northern Finland."[18] Along with this territorial and geographic fragmentation, the lack of a constitutional legal regime in the region further complicates the situation in the Arctic. With its complicated geographical situation, activity in the Arctic has never been limited to peaceful purposes; in contrast to the Antarctic, the Arctic has been a highly militarized zone, especially during the Cold War when it was a frontline of the competition between the two superpowers.

The international legal situation in the Arctic also contrasts strongly with that in the continent of Antarctica, as described by Jacek Machowski: "the Arctic does not have a general treaty dealing with the region as a whole," which "makes the definition of the status of the Arctic under international law and the determination there of national sovereignty and jurisdiction an extremely difficult task."[19] At present, the legal regime covering the Arctic region remains "scatter[ed]throughout the numerous bilateral and multilateral instruments" for certain areas or selected issues and is marked by ambiguity and contradiction, which breeds disputes in the region.[20]

There are two main legal instruments offering guidance to the international community for conduct in the Arctic: the United Nations Convention on the Law of the Sea (UNCLOS) and the Svalbard Treaty. As a comprehensive multilateral treaty, the UNCLOS is the polestar of maritime law and ostensibly limits state sovereignty. In the Arctic, it informs the process of allocating newly available Arctic resources. The convention thus not only establishes general normative standards of conduct but also governs the extent of national maritime sovereignties.[21] It defines the rights of Arctic Circle countries in territorial sea and contiguous zones, exclusive economic zones, and continental shelf

---

[18] Roach, J. A. (2020). *Freedom of the seas in the Arctic region.* In K. Spohr, D. S. Hamilton, & J. C. Moyer (Eds.), *The Arctic and world order* (pp. 219–250). Brookings.

[19] Machowski, J. (1995). Scientific activities on Spitsbergen in the light of the international legal status of the archipelago. *Polish Polar Research, 16*(1/2), 13–35.

[20] Ibid.

[21] Clote, P. (2008). Implications of global warming on state sovereignty and Arctic resources under the United Nations Convention on the Law of the Sea: How the Arctic is no longer communis omnium naturali jure. *Richmond Journal of Global Law and Business, 8*, 195.

areas in the Arctic region. The United States is the only non-Arctic state that has not ratified the UNCLOS. Therefore, it cannot submit territorial claims of extended continental shelf (ECS) to the Commission on the Limits of the Continental Shelf (CLCS), as these claims are reserved for member states.[22] The other main legal instrument is the Svalbard Treaty, which emerged from the 1920 Spitsbergen Treaty and offers an international legal framework only for a particular archipelago in the north of Norway.

2.3   MAJOR DISPUTES IN THE ARCTIC

As the result of legal vagueness and diverging interests, there remain a series of disputes among the member states of the Arctic Circle and between Arctic and non-Arctic countries. These disputes have escalated amid the geopolitical rivalries of recent decades. According to the analysis of Li Xueping,[23] Arctic and non-Arctic states have different emphases in relation to Arctic affairs. The Arctic states tend to focus on the following aspects: (1) jurisdiction over the Arctic passages, (2) sovereignty over continental shelf areas in the Arctic Ocean, (3) sharing of natural resources, and (4) environmental protection. The non-Arctic countries are more likely to assert their rights to (1) scientific research, (2) the high sea and international seabed, (3) innocent passage, and (4) resource exploitation and environmental protection.

To summarize, the major disputes in the Arctic region relate to (1) territorial claims, (2) the rights to passage and jurisdiction over Arctic sea routes, (3) the extension of the continental shelf areas in the Arctic Ocean, and (4) the right to conduct scientific research.

2.3.1   *Territorial Disputes*

Territorial disputes between Arctic states have been constantly exacerbated by intensifying climate change and the increasing accessibility of the Arctic. After a marathon negotiation, Russia and Norway finally resolved

[22] Frey, L. (2018). *Who owns the North Pole? Analysis of the territorial claims made by the Arctic states over the Continental Shelf in the Arctic Ocean.*

[23] Li, X. P., Fang, Z., & Liu, H. Y. (2017). Rights and obligations of the Arctic Region under the United Nations Convention on the Law of the Sea and their relationship with geographically disadvantaged States. *Polar Research 02*, 279–285.

their maritime border dispute in the Barents Sea by a delimitation agreement 10 years ago. However, there are several territorial disputes still active in the Arctic region, over (1) Hans Island between Denmark and Canada, (2) the Beaufort Sea between Canada and the United States, (3) the Northwest Passage, and (4) the Northeast Passage. We discuss the first two of these here, leaving the issue of the Northwest and Northeast Passages for the next part of the chapter.

1. *The Hans Island dispute between Denmark and Canada*

Hans Island is the last unsettled land in the Arctic. Situated in the Nares Strait, which runs between a northern island of Canada (Ellesmere Island) and Greenland, Hans Island is a small uninhabited island of only 1.3 square kilometers. In the 1970s, Canada and Denmark negotiated a treaty on the geographic coordinates of the continental shelf between the two countries, which has been in force since March 13, 1974. The two countries discovered that the island sat precisely in the middle of the line then being drawn to divide the continental shelf and maritime boundary between Canada and Greenland.[24] Although the treaty settled 127 points from Davis Strait to the end of Robeson Channel and defined the border between the two countries, Hans Island still hangs in the balance. In recent decades, Hans Island has shifted from "a legal point of contention between Canada and Denmark" to "a political one," with "the sovereignty over the small rock making more waves than its small stature could justify."[25]

In 1984, Demark planted a Danish flag on the island, which triggered a protest from Canada. In 2005, the then Canadian Defense Minister stepped on the island, fueling another diplomatic quarrel between the two countries, and a Danish naval vessel was dispatched toward the island. In 2008, a joint Automated Weather Station (AWS) was installed between the flagpoles, but as far as sovereignty is concerned, there have been no advances for more than a decade. In 2012, Canada and Denmark agreed on the exact border in the waters between Canada and Greenland all the way to the shores of Hans Island, but the island itself remains as disputed

[24] Gift, A. H. (2018). *Hans Island.*
[25] Ibid.

as ever.[26] In May 2018, the two countries announced that a joint task force had been set up to settle the dispute.

2. *The Beaufort Sea disputes between Canada and the United States*

The Beaufort Sea is a shallow area of the Arctic Ocean located between Alaska and the Canadian Arctic Archipelago, just to the north of the Mackenzie River delta.[27] In the Beaufort Sea, there is a wedge-shaped slice on the International Boundary that is the root of ongoing disputes between Canada and the United States. The dispute can be traced back to the Treaty of Saint Petersburg in 1825, which was meant to settle the land boundary between the United Kingdom and the Russian empire.[28] However, these territories ultimately became Yukon (belonging to Canada) and Alaska (belonging to the United States). The treaty could not pertain to the maritime boundary because in 1825 international law had not yet advanced to provide guidance beyond three nautical miles off the coastline.[29]

There are a range of factors in the ongoing disputes between Canada and the United States over the Beaufort Sea. According to the analysis of Rob Huebert, the geographic location of the Beaufort Sea involves it in a range of difficult issues[30]: (1) the determination of the maritime boundary between the United States and Canada, (2) the determination of the international legal status of the Northwest Passage, and (3) the formation of the boundary delimiting the area contained within the extended continental shelf off the coast of Alaska and the western edge of the Canadian Arctic. The large resource deposits in the region also increase the difficulty of settling the dispute: as far back as the

[26] Breum, M. (2018). Analysis: Hans Island and the endless dispute over its sovereignty. *High North News.* https://www.highnorthnews.com/en/analysis-hans-island-and-endless-dispute-over-its-sovereignty.

[27] Baker, J. S., & Byers, M. (2012). Crossed lines: The curious case of the Beaufort Sea maritime boundary dispute. *Ocean Development & International Law, 43*(1), 70–95.

[28] Gray, D. (1997). Canada's unresolved maritime boundaries. *IBRU Boundary and Security Bulletin.*

[29] Huebert, R. (2018). Drawing boundaries in the Beaufort Sea: Different visions/different needs. *Journal of Borderlands Studies, 33*(2), 203–223.

[30] Ibid.

1970s, seismic surveys and exploratory wells established the presence of hydrocarbons in seabed sediments.

Canada insists that the maritime boundary should extend the land boundary in a straight line and the United States argues that the maritime boundary should extend along a path equidistant from the coasts of the two nations. Given that the United States has not ratified the UNCLOS, the dispute on the maritime border is likely to be solved bilaterally. Neither side has been willing to compromise on this issue, but Baker and Byers pointed to some changes in recent years that may accelerate the resolution of this dispute: (1) the imperatives of the process for establishing sovereign rights over the extended continental shelf—Canada ratified the UNCLOS in 2003 and is supposed to make a submission to the CLCS by 2013; (2) rising oil prices and the inevitability of rising natural gas prices; and (3) the realization that the continental shelf areas in the Beaufort Sea probably extend much further than had been previously thought. Baker and Byers also argue that "Although discussions are being conducted behind closed doors, it is reasonable to surmise that the existing legal positions of the two countries will be a starting point."[31]

2.3.2 *The Rights of Passage on and Jurisdiction Over the Arctic Sea Routes*

The possibility of opening the Arctic passage is increasing with the intensification of global warming. In the competition for passage, Russia and Canada are geographically advantaged countries in terms of their large territory, and both claim exclusive rights when it comes to the Arctic passage. This triggers international concerns about the rights of passage on and jurisdiction over Arctic routes.

The issue mostly surrounds the Northeast Passage and Northwest Passage, both of which are sea routes through the Arctic Ocean. The former is between the Pacific and Atlantic Oceans, along the coast of Russia and Norway, and the latter is along the coast of North America via the waterways through the Canadian Arctic Archipelago. Russia claims that the Northeast Passage passes through its territorial waters in the Karam Vilkitsky and Sannikov Strait, and Canada claims that the Northwest Passage is a part of its internal waters.

---

[31] Baker, J. S., & Byers, M. (2012). Crossed lines: The curious case of the Beaufort Sea maritime boundary dispute. *Ocean Development & International Law, 43*(1), 70–95.

Russia and Canada put forward their claims based on a simple appeal to Article 7 of the UNCLOS,[32] which clearly defines the code of conduct in the internal waters and territorial sea of coastal countries. According to the UNCLOS, the internal waters are under the sovereignty of the coastal states; therefore, foreign vessels and aircraft have no right of passage through or over internal waters without the authority of the coastal state.[33] Similarly, territorial sea is under the sovereignty of the coastal state, which precludes innocent passage for aircraft in the airspace over the territorial sea.[34]

Russia and Canada have also bolstered their claims by establishing domestic laws. Canada's Ocean Act of 1996 asserts that "all waters within the bounds of the Canadian Arctic islands, including the Northwest Passage, are within its internal waters." In 1998, the State Duma of the Russian Federation adopted the Federation Law on Inland Waters, Territorial Sea and Contiguous Zone, which includes the following:

> Navigation on the waterways of the Northern Sea Route, the historical national unified transport line of communication of the Russian Federation in the Arctic, including the Vilkitsky, Shokalskiy, Dmitry Laptev and Sannikov straits, shall be carried out in accordance with this Federal Act, other federal laws and the international treaties to which the Russian Federation is a party and the regulations on navigation on the watercourses of the Northern Sea Route approved by the Government of the Russian Federation and published in Notices to Mariners.[35]

If the claims of Russia and Canada were to be recognized, the movement of other countries would be severely limited by the great proportion of territory that these two countries would control in the region. Accordingly, the international community has responded to the claims of Russia and Canada with strong opposition and an insistence on their rights of innocent passage according to the UNCLOS.

---

[32] United Nations Convention on the Law of the Sea (UNCLOS).

[33] UNCLOS, Article 2(1–2).

[34] UNCLOS, Articles 2(2) & 17–20. The rights and duties of the coastal state are set out in Articles 24–26.

[35] Federal Act on the Internal Maritime Waters, Territorial Sea and Contiguous Zone of the Russian Federation.

The United States is also a dominant power in the Arctic region and, despite its alliance with Canada, its attitude to the passage jurisdiction differs from those of Russia and Canada. The United States and the European Union consistently argue that both the Northeast and Northwest Passages should be defined as international waters and transit passages, considering the crucial importance to international navigation of the routes through the Arctic. The United States also strongly emphasizes "freedom of navigation and overflight" in the region. There are two reasons for the United States to hold this position: First, the Arctic territory of the United States is not as large as that of Russia or Canada; second, "the freedom of navigation and overflight" accords with its global strategy and helps the United States to maintain its worldwide military dominance. As Fahey[36] pointed out, "Ensuring freedom of the seas globally has long been a central component of U.S. national security policy and preserving freedom of the seas is a key objective of the U.S. National Strategy for the Arctic Region." According to the definition provided by the United States Department of Defense, the claims of Russia and Canada contradict the notion of "freedom of the seas":

> Freedom of the seas, however, includes more than the mere freedom of commercial vessels to transit through international waterways. While not a defined term under international law, the Department uses "freedom of the seas" to mean all of the rights, freedoms, and lawful uses of the sea and airspace, including for military ships and aircraft, recognized under international law. Freedom of the seas is thus also essential to ensure access in the event of a crisis. Conflicts and disasters can threaten U.S. interests and those of our regional allies and partners. The Department of Defense is therefore committed to ensuring free and open maritime access to protect the stable economic order that has served all Asia-Pacific nations so well for so long, and to maintain the ability of U.S. forces to respond as needed.[37]

There is therefore a fundamental divergence between the United States and the other non-Arctic countries on the issue of rights of passage on the maritime routes through the Arctic Ocean.

---

[36] Fahey, S. (2018). Access control: Freedom of the seas in the Arctic and the Russian Northern Sea Route regime. *Harvard National Security Journal, 9*, 154.

[37] United States Department of Defense. (2015). *The Asia–Pacific maritime security strategy: Achieving US national security objectives in a changing environment.*

2.3.3   Extending the Continental Shelf and the "Common Heritage of Mankind"

With intensifying global warming and technological advancements, recent decades have witnessed an increasing number of claims to the continental shelf areas in the Arctic region. These claims arise from interests in the extensive resources of the region and the accelerated opening of new maritime routes as a result of global warming. Against this background, delimiting the continental shelf areas is another crucial element of territorial disputes among Arctic Circle countries. According to the definition of the CLCS,

> The term "continental shelf" is used by geologists generally to mean that part of the continental margin which is between the shoreline and the shelf break or, where there is no noticeable slope, between the shoreline and the point where the depth of the superjacent water is approximately between 100 and 200 meters. However, this term is used in article 76 as a juridical term.[38]

In Article 76 of the UNCLOS, "continental shelf" is defined as follows:

> The continental shelf of a coastal State comprises the seabed and subsoil of the submarine areas that extend beyond its territorial sea throughout the natural prolongation of its land territory to the outer edge of the continental margin, or to a distance of 200 nautical miles from the baselines from which the breadth of the territorial sea is measured where the outer edge of the continental margin does not extend up to that distance.[39]

However, a series of problems remain in determining the international legal status of the continental shelf, which escalates disputes between Arctic Circle countries. Further rivalries are triggered by the declaration under the UNCLOS that a coastal state may "claim an extended continental shelf up to 350 nautical miles from its baselines by proving that this area is a natural prolongation of the state's land territory."

As the largest Arctic country, Russia is the most active party in making claims to the extension of the continental shelf and the relevant sovereign rights up to the North Pole, and is particularly active in claiming sovereign

---

[38] Commission on the Limits of the Continental Shelf (CLCS).
[39] UNCLOS.

rights over vast parts of the Arctic Ocean floor. Two significant events in recent decades have indicated Russia's resolve: The first was its formal submission to the CLCS to extend the continental shelf in 2001, and the second was the symbolic planting of a Russian flag on the ocean floor during a scientific expedition in 2007.[40] Matz-Luck pointed out that the submission to the CLCS is "relevant for the legal proceedings in the determination of Russian claims," whereas planting the flag is of "only political relevance."[41]

Kuznetsova and Loshchakov (2021)[42] argued that

> Posture questions of the Russian Federation in relation to the Arctic are determined, among other things, by the proximity of the Kola Peninsula (Murmansk region, Murmansk) to the Arctic region; the existence of an ice-free deep-water port, the world's only base of atomic icebreakers, as well as transport, fishing, military and research fleets. Strategic vector of Russia is related to the reclamation of the Arctic shelf through effective technical and technological support of the Northern Sea Route and the development of the transport and logistics infrastructure of the Arctic. Therefore, the introduction of the legal aspects in the regulation of existing dispute at the international level on the issues of the continental Arctic shelf is one of the main aims of the Russian Federation in the execution of its intentions.

However, the unilateral actions of Russia on the Arctic continental shelf have triggered a chorus of opposition from the international community, because its claims severely challenge the rights of the other countries in the Arctic, especially those who are not geographically proximate to the region. It is also said to pose a challenge to the "common heritage of mankind" (CHM), also known as "the common heritage of humankind or humanity," which represents the notion that "certain global commons or elements regarded as beneficial to humanity as a whole should not be unilaterally exploited by individual states or their nationals, nor by corporations or other entities, but rather should be exploited under some

---

[40] Matz-Luck, N. (2009). Planting a flag in Arctic waters: Russia's claim to the North Pole. *Goettingen Journal of International Law*, *1*, 235.

[41] Ibid.

[42] Kuznetsova, E., & Loshchakov, D. (2021). Actual legal situation and international controversy concerning the continental shelf in the Arctic zone. *Laplage em Revista*, *7*(Extra-E), 584–590.

sort of international arrangement or regime for the benefit of mankind as a whole."[43]

According to the UNCLOS, in areas that are considered as the CHM,

> No State shall claim or exercise sovereignty or sovereign rights over any part of the Area or its resources, nor shall any State or natural or juridical person appropriate any part thereof. No such claim or exercise of sovereignty or sovereign rights nor such appropriation shall be recognized.[44]

These areas over which national jurisdiction is limited include the seabed, the ocean floor and subsoil, and the resources that they hold. In such an area, "all rights in the resources ... are vested in mankind as a whole"[45] and "exploration and exploitation ... shall be carried out for the benefit of mankind as a whole, irrespective of the geographical location of States."[46] As discussed in the previous section, some Arctic countries are seeking the extension of their continental shelf to give them the rights to its exploitation. These claims present a range of challenges to the rights of non-Arctic countries over these areas of the CHM.

2.3.4   The Right to Scientific Research

Polar research has experienced rapid development in recent decades marked by climate change. Countries that have ambitions to develop polar research seek rights of access to the polar regions. In the Arctic, there are two main legal instruments facilitating international scientific research: the UNCLOS and the Svalbard Treaty.

The UNCLOS defines the right to conduct marine scientific research in Article 238, Part XIII:

> All States, irrespective of their geographical location, and competent international organizations have the right to conduct marine scientific research subject to the rights and duties of other States as provided for in this Convention.

---

[43] Egede, E. (2017). *Common heritage of mankind.*

[44] UNCLOS, Article 137(1).

[45] UNCLOS, Article 137(2).

[46] UNCLOS, Preamble.

Based on this article, non-Arctic states can enjoy absolute rights of scientific research without authorization except for regions under the sovereignty of Arctic countries, such as continental shelf and exclusive economic zones.

The Svalbard Treaty is another pillar supporting international scientific research in the Arctic. Also known as Spitzbergen, Svalbard is a Norwegian archipelago in the Arctic Ocean midway between the northern coast of Norway and the North Pole. Article 5 of the Svalbard Treaty defines the right to scientific research in the Norwegian archipelago:

> The High Contracting Parties recognize the utility of establishing an international meteorological station in the territories specified in Article 1, the organization of which shall form the subject of a subsequent Convention. ... Conventions shall also be concluded laying down the conditions under which scientific investigations may be conducted in the said territories.[47]

For decades, the archipelago has served as "a large arctic natural laboratory," occupying a special position in polar science and international cooperation.[48] However, because of the small size of the Spitzbergen Archipelago, the scientific activities of the non-Arctic states in the Arctic as a whole still face a series of challenges.

2.4 Summary

This chapter provides a panoramic view of the rapidly changing Arctic, especially from the perspectives of its legal status, disputes, human civilization, and heritage. As a frozen land that has experienced the effects of rapid climate change over the past few decades, the Arctic has risen to become a new focus for competition between and among states, despite being covered with snow and ice throughout the year. Multilateralism provides a theoretical framework for observing and analyzing the competition and cooperation of major countries in the Arctic, especially with the formation of a new strategic triangle comprising China, the United States, and Russia.

---

[47] The Svalbard Treaty. (1920).

[48] Machowski, J. (1995). Scientific activities on Spitsbergen in the light of the international legal status of the archipelago. *Polish Polar Research, 16*(1/2), 13–35

REFERENCES

Baker, J. S., & Byers, M. (2012). Crossed lines: The curious case of the Beaufort Sea maritime boundary dispute. *Ocean Development & International Law, 43*(1), 70–95.

Bjola, C. (2013). Keeping the Arctic "cold": The rise of plurilateral diplomacy? *Global Policy, 4*(4), 347–358.

Breum, M. (2018). *Analysis: Hans Island and the endless dispute over its sovereignty.* High North News. https://www.highnorthnews.com/en/ana lysis-hans-island-and-endless-dispute-over-its-sovereignty

Clote, P. (2008). Implications of global warming on state sovereignty and Arctic resources under the United Nations Convention on the Law of the Sea: How the Arctic is no longer communis omnium naturali jure. *Richmond Journal of Global Law and Business, 8*(2), 195–247.

Egede, E. (2017). *Common heritage of mankind.* https://www.oxfordbiblio graphies.com/view/document/obo-9780199796953/obo-9780199796953-0109.xml

Exner-Pirot, H. (2013). What is the Arctic a case of? The Arctic as a regional environmental security complex and the implications for policy. *The Polar Journal, 3*(1), 120–135.

Fahey, S. (2018). Access control: Freedom of the seas in the Arctic and the Russian Northern Sea Route regime. *Harvard National Security Journal, 9*, 154–200.

Federal Act on the Internal Maritime Waters, Territorial Sea and Contiguous Zone of the Russian Federation. https://www.un.org/depts/los/LEGISL ATIONANDTREATIES/PDFFILES/RUS_1998_Act_TS.pdf

Frey, L. (2018). *Who owns the North Pole? Analysis of the territorial claims made by the Arctic states over the Continental Shelf in the Arctic Ocean.* https://osuva.uwasa.fi/handle/10024/9655

Gift, A. H. (2018). *Hans Island.* http://adamlajeunesse.com/uploads/3/4/9/1/34912685/hans_island_-_3.pdf

Gray, D. (1997). *Canada's unresolved maritime boundaries.* IBRU Boundary and Security Bulletin. https://www.dur.ac.uk/resources/ibru/publications/full/bsb5-3_gray.pdf

Heininen, L. (2012). *State of the Arctic strategies and policies: A summary.* Arctic Yearbook 2012. https://arcticyearbook.com/images/Articles_2017/scholarly-articles/7_Urban_Planning_in_the_Arctic.pdf

Huebert, R. (2018). Drawing boundaries in the Beaufort Sea: Different visions/different needs. *Journal of Borderlands Studies, 33*(2), 203–223.

Kuznetsova, E., & Loshchakov, D. (2021). Actual legal situation and international controversy concerning the continental shelf in the Arctic zone. *Laplage em Revista, 7*(Extra-E), 584–590.

Li, X. P., Fang, Z., & Liu, H. Y. (2017). Rights and obligations of the Arctic Region under the United Nations Convention on the Law of the Sea and their relationship with geographically disadvantaged States. *Polar Research, 02*, 279–285.

Machowski, J. (1995). Scientific activities on Spitsbergen in the light of the international legal status of the archipelago. *Polish Polar Research, 16*(1/2), 13–35.

Matz-Luck, N. (2009). Planting a flag in Arctic waters: Russia's claim to the North Pole. *Goettingen Journal of International Law, 1*, 235–256.

NSIDC. (2008). *Sea ice decline accelerates, Amundsen's Northwest Passage opens*. https://nsidc.org/arcticseaicenews/2008/08/sea-ice-decline-accelerates-amundsens-northwest-passage-opens/

Roach, J. A. (2020). Freedom of the seas in the Arctic region. In K. Spohr, D. S. Hamilton, & J. C. Moyer (Eds.), *The Arctic and world order* (pp. 219–250). Brookings Institution Press.

Sheng, L. (2012). Dealing with financial risks of international capital flows: A theoretical framework. *Cambridge Review of International Affairs, 25*(3), 463–474.

Sheng, L. (2014). Capital controls and international development: A theoretical reconsideration. *Global Policy, 5*(1), 114–120.

Sheng, L. (2014). Economic structure, cost outsourcing and global imbalances. *Journal of Australian Political Economy, 74*, 81–94.

Sheng, L. (2015). Theorizing income inequality in the face of financial globalization. *The Social Science Journal, 52*(3), 415–424.

Sheng, L. (2021). *How Covid-19 reshapes new world order: Political economy perspective*. Springer.

Sheng, L., & do Nascimento, D. F. (2021). *Love and trade war: China and the US in historical context*. Springer.

Tamnes, R., & Holtsmark, S. G. (2014). The geopolitics of the Arctic in historical perspective. In R. Tamnes & K. Offerdal (Eds.), *Geopolitics and security in the Arctic* (pp. 26–62). Routledge.

The Antarctic Treaty. (1959). https://www.bas.ac.uk/about/antarctica/the-antarctic-treaty/the-antarctic-treaty-1959/

The Svalbard Treaty. (1920). http://library.arcticportal.org/1909/1/The_Svalbard_Treaty_9ssFy.pdf

United Nations Convention on the Law of the Sea (UNCLOS). https://www.un.org/depts/los/convention_agreements/texts/unclos/unclos_e.pdf

United States Department of Defense. (2015). *The Asia-Pacific maritime security strategy: Achieving US national security objectives in a changing environment*. https://dod.defense.gov/Portals/1/Documents/pubs/NDAA%20A-P_Maritime_SecuritY_Strategy-08142015-1300-FINALFORMAT.PDF

Wegge, N. (2011). The political order in the Arctic: Power structures, regimes and influence. *Polar Record, 47*(2), 165–176.

Zellen, B. S. (2009). *Arctic doom, Arctic boom: The geopolitics of climate change in the Arctic.* https://www.researchgate.net/publication/309650349_Arctic_Doom_Arctic_Boom_The_Geopolitics_of_Climate_Change_in_the_Arctic

CHAPTER 3

The New Strategic Triangle in the Arctic: China, Russia, and the United States

Abstract The Arctic is becoming an important arena of contestation between the great powers in the context of global warming. As China has an increasingly strong presence in the region, the bipolar competition in the Arctic between Russia and the United States is shifting into a triangular game. This chapter elaborates on the formation of a new strategic triangle between China, Russia, and the United States and the gradual change in the balance of power in the Arctic as it becomes a hotspot for interactions between the three powers.

Keywords Great Triangle Theory · China–US–Russia triangle · World order · Great power competition

China, Russia, and the United States are the three most vital players in the international arena, and their interactions have deeply shaped the landscape of world politics since the end of World War II. During the Cold War, China was a game changer in the bipolar system dominated by the United States and the Soviet Union. In the 1970s, the notion of "strategic triangle" was formed and originally indicated the trilateral relations between China, the Soviet Union, and the United States. Even today, after the collapse of the Soviet Union, these three powers still play

© The Author(s), under exclusive license to Springer Nature
Singapore Pte Ltd. 2022
E. L. Sheng, *Arctic Opportunities and Challenges*,
https://doi.org/10.1007/978-981-19-1246-7_3

important roles in shaping global geopolitics, and their trilateral relations are still the most significant of those among the major powers.

There are still heated debates over the notion of a strategic triangle involving China, Russia, and the United States. The focus of these debates is usually on the comparison between the triangular relations of the 1970s and those prevailing today. It is believed, the current triangle is the successor of that of the Cold War because of its highly similar structure. Three elements of this similarity are highlighted: (1) there is an absolutely dominant player; (2) the relations between the three parties are highly interactive; and (3) the weakest party in the triangle is likely to decide the balance of power of the whole structure. However, others have insisted that it is too early to discuss this topic, because "Washington [has] refused to acknowledge its existence as a matter of policy concern," despite China and Russia forming a strategic alliance while Washington considers both as strategic competitors from which it must not shirk.[1,2]

In this chapter, we argue that the dynamics of the Arctic epitomize the formation of a new strategic triangle of China, Russia, and the United States. Like the fulcrum of a seesaw, the dynamic game of Arctic affairs props up this new strategic triangle. The game between these three countries in different hemispheres thus has a new playing field. The participation of these three countries in Arctic affairs will determine their roles in this game and the future trends of Arctic geopolitics. Similarly, the development of Arctic affairs will bring innovations to the interaction among the three countries and thus have an impact on the formation and transformation of the strategic triangle.

3.1   A New Strategic Triangle? Cyclical History and Recent Developments

The Arctic has been dominated by the United States and Russia for half a century. However, China's increasing presence over the past decade has changed the nature of the game. Benefiting from its geoeconomic strategy

[1] Blank, S. (2019). Triangularism old and new: China, Russia, and the United States. In J. I. Bekkevold & B. Lo (Eds.), *Sino-Russian relations in the 21st Century* (pp. 215–241). Springer.

[2] Graham, T. (2020). China-Russia-US relations and strategic triangles. *Political Studies, 6*(6), 62–72.

to promote export-oriented industrialization, China have enjoyed impressive economic growth over the past few decades at an annual growth rate over 6% and become the world's second-largest economy.[3,4] As an extra-regional actor in the Arctic, China's presence in the region has increased in two primary ways: first, by participating in the Arctic Council as an observer state; and second, through enhanced bilateral and multilateral cooperation with Arctic states, especially Russia. Since the early 2000s, China, like many other non-regional states, increasingly recognized the gatekeeping role of the Arctic Council.[5] In recent years, Sino–Russian cooperation has been emphasized against the background of Sino–American tensions.[6] The Arctic cooperation between China and Russia has developed rapidly within the framework of the Polar Silk Road and BRI, but these developments are raising alarms in the United States due to security concerns.[7] It is believed that the Arctic will become the new arena of Sino–American competition with the formation of a strategic triangle and that it can thus be seen as the epitome of the trilateral power game. For example, Rob Huebert described the new situation in the Arctic as the "New Arctic Strategic Triangle Environment."[8]

Given the increasing intricacy of international relations, there is cause to wonder how the situation in the Arctic will evolve. Three questions are central to these considerations. First, how will the three powers—China, Russia, and the United States—interact in the region and has a new strategic triangle already been formed? Second, what are the similarities and differences between this new strategic triangle and the strategic

[3] Sheng, L. (2014b). Economic structure, cost outsourcing and global imbalances. *Journal of Australian Political Economy, 74*, 81–94.

[4] Sheng, L. (2016). Explaining US-China economic imbalances: A social perspective. *Cambridge Review of International Affairs, 29*(3), 1097–1111.

[5] Stephen, M. D., & Stephen, K. (2020). The integration of emerging powers into club institutions: China and the Arctic Council. *Global Policy, 11*, 51–60.

[6] Sheng, L., & Nascimento, D. F. (2021). *The Belt and road initiative in South-South Cooperation: The impact on world trade and geopolitics*. Palgrave Macmillan.

[7] Allison, G. (2018, December 14). *China and Russia: A strategic alliance in the making*. The National Interest; The Center for the National Interest.

[8] Huebert, R. (2019). The New Arctic Strategic Triangle Environment (NASTE). *Breaking the ice curtain? Russia, Canada, and Arctic security in a changing circumpolar world, Canadian Global Affairs Institute report* (pp. 75–93). https://d3n8a8pro7vhmx. cloudfront.net/cdfai/pages/4193/attachments/original/1558816637/Breaking_the_Ice_ Curtain.pdf?1558816637.

triangle as it is conceived in traditional international relations? Third, how are we to observe and measure the influence of this new strategic triangle on competition and cooperation between the three great powers and on the present and future situation in the Arctic region? This chapter revisits the interaction of these three powers in the historical period of the 1970s when the China–Soviet Union–United States strategic triangle was in place. This discussion helps to present a more comprehensive and profound snapshot of the strategic diplomacy and multilateral relations between China, the United States, and Russia today. Accordingly, the chapter segues into further discussion of unequal interactive patterns of cooperation and competition in the new triangular relationship and the configuration of, and trends in the development of, the new strategic triangle in the Arctic.

The vicissitudes of world history always revolve around the rise and fall of great powers. The disputes and interest gambling among great powers and the consequent impact on the global and regional situation promote, restrict, or even halt the development of world history. In different historical stages, the actions and interactions of great powers differ in their methods and goals and determine global patterns.[9] The end of the Cold War broke the pattern of a bipolar world dominated by the confrontation between the United States and the Soviet Union, which had lasted for nearly half a century. It was replaced by a new world order characterized by multipolarization,[10] within which the role and status of the relations between great powers have significantly increased. The adjustment of their respective interests, repositioning of their mutual relations to the new situation, and reconstruction of their foreign policies and attitudes and views on international affairs all affect the stability and development of the whole world.[11]

Accordingly, as one of the most closely watched multilateral relations, the change and development of Sino–American–Russian relations not only affects the domestic political, economic, and social development of the three parties but also has a certain influence on the patterns and future

[9] Sheng, L. (2021). *How Covid-19 reshapes new world order: Political economy perspective.* Springer.

[10] Gaddis, J. L. (1992). International relations theory and the end of the Cold War. *International Security, 17*(3), 5–58.

[11] Cui, S., & Buzan, B. (2016). Great power management in international society. *The Chinese Journal of International Politics, 9*(2), 181–210.

developments of a new multipolar world and on the new world order as such. The participation of these three powers in Arctic affairs, and the decisions they make in this context, is no exception. Therefore, looking back and examining the strategic triangular relationship between China, the United States, and Russia can improve our understanding and analysis of their interaction in Arctic affairs and the likely future trends.

3.1.1   The Genesis and Decay of the United States–China–Soviet Union Strategic Triangle

The trilateral relations between China, Russia, and the United States in the Cold War are widely known as "the strategic triangle." From the mid-1950s, the relations between Beijing and Moscow rapidly deteriorated as the result of historical rows over ideological divergences and different national interests. The confrontations between China and the Soviet Union worsened during the 1960s,[12] and constant friction on the border of China and the Soviet Union eventually escalated into direct conflict on Zhenbao Island in 1969. Both sides claimed that the other had fired first, and although both declared victory, the battle killed hundreds of soldiers on both sides without a decisive outcome.

The U.S. Central Intelligence Agency (CIA) had in fact detected the escalating confrontations between China and the Soviet Union and rapid changes in the buildup of military power on the border of the two countries earlier in the same year. As tensions between Moscow and Beijing held potential strategic benefits for Washington, President Nixon decided to re-evaluate the military and foreign polices of China and the Soviet Union in an effort to ascertain whether the United States could take the opportunity offered by Sino-Soviet conflict to expand its strategic space.[13] The preliminary conclusion brought to the Nixon administration from this assessment was that it was unlikely that China and the Soviet Union would repair their relations in the short term. Although China attached strategic importance to both the United States and the Soviet Union, the latter would endeavor to curb the construction of Sino-American relations. Meanwhile, China's influence in Asia was assessed as political rather

---

[12] Holmes, C. (1979). The Soviet Union and China. In *The Soviet Union in East Asia*. Routledge.

[13] Kissinger, H. A. (1969). *National security study memorandum 63*. United States.

than military. In the case of military conflict between China and the Soviet Union, China would not confront the United States on a new front.[14]

China also observed changes in the international situation and planned to adjust its relations with the Soviet Union and the United States accordingly. The direct military threat of the United States to China was weakening while that of the Soviet Union was increasing day by day. Nixon and Kissinger began to consider how to take advantage of this opportunity to enhance their initiative. They tried to stimulate and guide the two countries to seek to improve their relations with the United States to constitute a strategic triangle, which would first require the arduous task of establishing diplomatic relations with China and building Sino-American ties.[15]

To a large extent, Kissinger accomplished this feat. From the perspective of the United States, triangular diplomacy was a huge success. The American foreign policy of the time also enabled the Soviet Union and China to jointly advance their respective strategic interests. With the relaxation of tensions, the advantages of the Soviet government in geopolitics were recognized. From China's point of view, the confrontation with the Soviet Union had given it stronger security and initiative on issues related to Taiwan. Although the United States gained the most from this game, the two other parties also received payback, which is a key reason for the strategic triangle's initial emergence.[16]

To summarize, the strategic triangle emerged in 1969 out of dramatic changes in the relations between China, the United States, and the Soviet Union. China and the United States began to re-establish ties and sought to establish and expand their consensus over certain strategies. China thereby became an important anchor point in the trilateral relations. For China, this marked the abandonment of its formerly one-sided diplomatic approach with the Soviet Union and the beginnings of diplomacy with the United States and the West. For the United States, this marked the beginning of a new strategic relationship with China. For the Soviet Union, however, its failure to prevent the establishment of diplomatic relations

[14] Feus. (1969). *Foreign relations of the United States, 1969–1976, Volume XVII, China, 1969–1972.*

[15] Kissinger, H. (2011). *White house years.* Simon and Schuster.

[16] Ibid.

between China and the United States meant that it began to play a more passive role in the tripartite relations.[17]

The situation changed when President Carter took office in the United States in 1977. The Carter administration did not adhere to the existing triangular relations by seeking to improve its relations with Moscow and Beijing. Instead, it moved closer to China to put pressure on the Soviet Union to make concessions to the American position. This led to the end of the detente relationship supported by the strategic triangle between China, the Soviet Union, and the United States. In the 1980s, the importance of the strategic triangle was diluted further; with the disintegration of the Soviet Union in 1991, the triangular relationship was no longer effective for any of the three actors and its influence waned.[18]

3.1.2 Is a New Strategic Triangle Forming?

Since the end of the Cold War, there have been significant changes in international relations. The end of the American–Soviet bipolar pattern led the world to a crossroads, with the promotion of the development of a pattern of multipolarization as the realistic basis for building a new world order.[19] At the same time, rapid globalization has not only deepened the interdependence of countries across the globe but also intensified their competition, especially since the dawn of the new millennium.[20,21,22] China's entry to the World Trade Organization (WTO) in 2001 not only accelerated its rise and strengthened its role in international affairs and world politics but also raised alarms in the West, particularly because of its threat to the hegemony of the United States. According to the analysis

[17] Gaddis, J. L., & Bothwell, R. (1997). We now know: Rethinking cold-war history. *International Journal, 52*(3), 537–560.

[18] Ross, R. S. (2015). *China, the United States, and the Soviet Union: Tripolarity and policy making in the Cold War.* M. E. Sharpe.

[19] Varisco, A. E. (2013). Towards a multi-polar international system: Which prospects for global peace? *E-International Relations Students, 3.*

[20] Sheng, L. (2012). Dealing with financial risks of international capital flows: A theoretical framework. *Cambridge Review of International Affairs, 25*(3), 463–474.

[21] Sheng, L. (2014a). Capital controls and international development: A theoretical reconsideration. *Global Policy, 5*(1), 114–120.

[22] Sheng, L. (2017). Explaining urban economic governance: The city of Macao. *Cities, 61,* 96–108.

of Mastanduno,[23] the United States enlarged its hegemony in the immediate post-Cold War era by offering partnerships that differed significantly based on the relative importance of each state in the management of the American hegemonic order. That order is now in jeopardy because the United States failed to form a partnership with Russia and the partnership it did succeed in forming with China and maintained for two decades is now under considerable strain.

Tensions between the major powers flared up once again during the global financial crisis of 2008 and 2009. Strategic cooperation between Russia and China has been increased as the two parties try to challenge the dominant position of the United States in the international system. As two permanent members of the United Nations Security Council, China and Russia have improved their cooperation in various fields, especially bilateral trade and traditional and non-traditional security matters. The bilateral trade between China and Russia increased from around US$64 billion in 2015 to over US$110 billion in 2020.[24] China is important for Russia in the economic field as its largest trading partner.[25] In addition to economic matters, China and Russia have carried out extensive cooperation in military and regional security affairs, including joint military exercises and coordinated positions in the United Nations Security Council and forums of other international organizations.[26]

At present, China and Russia have highly overlapping interests and, to some extent, the two countries have formed a kind of strategic community over these shared interests. The foreign ministers of both countries have pledged firm mutual support on core interests. Lavrov, the Russian Foreign Minister, stated that Russia is willing to carry out close strategic coordination with China on international and regional issues, provide firm support on issues concerning the core interests of both sides, and jointly

[23] Mastanduno, M. (2019). Partner politics: Russia, China, and the challenge of extending US hegemony after the Cold War. *Security Studies, 28*(3), 479–504.

[24] Statista. (2021b). *Value of Russian trade in goods (export, import, and trade balance) with China from 2007 to 2020.*

[25] Oxford Analytica. (2021). *Russia-China economic imbalance will deepen.* Emerald Expert Briefings.

[26] Abdenur, A. E. (2017). Can the BRICS cooperate in international security? *International Organisations Research Journal, 12*(3), 73–93.

oppose world hegemonism. In contrast to these blooming bilateral relations between China and Russia, relations between the United States and both China and Russia have fluctuated in recent years.

Relations between Russia and the United States began to sour with the Crimean crisis in 2014 under the administration of Barack Obama, and have since fallen into a stalemate. In fact, it has been claimed that relations between Russia and the United States are now the worst they have been in recent decades. Contradictions in the positions of Moscow and Washington are often compared with those of the Cold War, especially over the crises in Ukraine and Crimea. According to the analysis of Sushentsov and Suchkov (2019),[27] confrontations between Russia and the United States are found across many fields, such as politics, economics, and information, and the impacts of these confrontations on bilateral cooperation are therefore broad. They also pointed out that the fluctuating relations between Russia and the United States are rooted in a psychological phenomenon known as "fundamental attribution error," which is "a tendency towards explaining the behavior and actions of other people by their bad qualities and one's own behavior by external circumstances." Against the background of a de facto political "civil war" within the American establishment, Russian "meddling" was set up as a target for blame to distract public attention in the United States. This has diverted the decision-making process and the outcomes of foreign policy in the White House. The activity of Russian hackers and Russia's suppression of dissidents play similar roles in the recent relations between the two countries. Meanwhile, Russia accuses the United States of promoting regime subversion and "color revolutions" against the Kremlin.[28] For these reasons, it is believed that relations between the United States and Russia will remain poor for the next few years at least, with a lingering risk of escalation.[29]

Relations between China and the United States have rapidly deteriorated since the Obama administration expressed its concerns over China's cyberespionage and military buildup in the South China Sea. When

---

[27] Sushentsov, A. A., & Suchkov, M. A. (2019). *The nature of the modern crisis in US-Russia relations*. Russia in Global Affairs (4).

[28] MacMillan, S., & Outlook, N. E. (2015). *Coups and "Color Revolutions": US wages geopolitical warfare against Russia in Central Asia and Caucasus*. Global Research, 9.

[29] Newlin, C., Conley, H. A., Viakhireva, N., & Timofeev, I. (2020). *US-Russia relations at a crossroads*. Center for Strategic & International Studies (CSIS).

Donald Trump took office in the United States, these relations plummeted into open rivalry spanning the economic, military, technological, and ideological spheres and resounding with echoes of the Cold War.[30] During the four years of Trump's administration from 2016 to 2020, guided by his "America First" philosophy, Trump and his colleagues made a series of aggressive moves against China in politics, economics, trade, and high technology. Donald Trump fulfilled his election promise by declaring a trade war on China in 2017, which had a systemic impact on the international political order and world economic structure.[31] The Trump administration also irritated Beijing by constantly pointing its fingers at China's domestic affairs, such as the status of Taiwan, the Hong Kong riots in 2019, and the anti-pandemic methods adopted by China.

Despite the election of a Democratic President in the 2020 campaign, there appears to have been little improvement in the Sino–American relations in 2021. Although American politics is currently marked by a sharp divergence between the Republicans and Democrats, Joe Biden has inherited the legacy of Donald Trump in foreign policy toward China, particularly in relation to trade.[32] According to Piesse,[33] major change on this front is unlikely for two reasons. First, the Biden administration is unlikely to adopt a significantly different trade policy vis-à-vis China than that of the previous administration: "Tariffs and other measures that encourage systemic change within China look set to continue." Second, although the US–China Phase One trade deal is under review, it is unlikely that the Biden administration will make significant alterations: "China is not expected to meet the import targets set in the trade deal, but US officials appear to be satisfied with the progress that Beijing has made under the agreement. Statements from several senior members of the Biden Administration also suggest that the process of economic decoupling from China is likely to continue in some form."

The triangular structure of China, Russia, and the United States is still clearly observable in the international arena, although the arrangements are quite different from those in the 1970s. As reviewed above, during

[30] Ibid.

[31] Sheng, L., & do Nascimento, D. F. (2021). *Love and trade war: China and the US in historical context.* Springer.

[32] Carr Jr, E. A. (2021). *From Trump to Biden and beyond: Reimagining US–China relations.* Springer.

[33] Piesse, M. (2021). *The US-China trade relationship during the Biden Administration.*

the Cold War, the Soviet Union was a significant threat to the interests of both the United States and China, which laid the foundation for China and the United States to draw closer and jointly cope with the threat. At that time, China was the weakest of the three countries but it nonetheless played an important role because China's preference for either side could change the balance of power in the confrontation between the United States and the Soviet Union. At present, the spearhead seems to point to the United States, and it is Russia and China that feel threatened in the emerging triangular relationship.

Eugene Rumer, director of Carnegie's Russia and Eurasia Program, and Richard Sokolsky, a researcher at the same center, pointed out that the foundation of the current Sino–Russian partnership is that the United States has made negative relations with China and Russia at the same time. Both China and Russia are attempting to curb American hegemony based on its pursuit of democracy as an instrument and unilateral use of force. They believe that after the Cold War, the United States, both in propaganda and practice, has criticized the lack of democracy in Russia and China while supporting authoritarian regimes abroad and promoting Western democratic values. This practice is regarded as an infringement and a threat to sovereignty by both China and Russia. Moreover, both Russia and China believe that the alliances and military presence of the United States close to their borders threaten their national security.[34]

Graham, a political scientist and former diplomat, pointed out that we have not seen the recovery of the strategic triangle, largely because Washington refuses to recognize its existence as a policy concern. On this account, although there is no doubt that the United States sees both China and Russia as strategic competitors, Sino–Russian cooperation is not recognized as a great challenge for the United States. In Graham's opinion, the United States has regarded Russia as a battered country since the end of the Cold War, viewing it more as troublesome than a major threat and seldom paying attention to affairs within Russia itself; Russia is merely a manageable challenge to the interests of the United States in Europe, the Middle East, and other regions.[35]

[34] Rumer, E., Sokolsky, R., & Weiss, A. S. (2017). Trump and Russia: The right way to manage relations. *Foreign Affairs, 96*, 12.

[35] Graham, T. (2020). China-Russia-US relations and strategic triangles. *Political Studies, 6*(6), 62–72.

Graham also argued that despite the constant pressure that both countries face from the United States, American policymakers doubt the durability of Sino–Russian strategic cooperation. American policymakers maintain the belief that the extent of any Sino–Russian strategic alliance is strictly limited and that China will be unwilling to play the role of a secondary partner for very long. There are fundamental differences between the design of the future world order sought by the two countries. Russia seems to be interested in overthrowing the current order and replacing it with the coordination of major powers based on spheres of influence; China, which has benefited from the current order, seems more inclined to reshape the balance in a favorable direction by being more heavily involved in the formulation of rules. American policymakers are convinced that as the importance of the United States to China's economic and security interests far exceeds that of Russia, China does not intend to sacrifice its complicated and delicate relationship with the United States to protect Russia. Furthermore, China has its own reservations about some Russian actions.[36]

As reviewed in the previous section, the United States once took advantage of the strategic opportunity to draw China over to the fight against the Soviet Union, which greatly influenced the playing field and the outcome of the Cold War. Based on Graham's observation, however, there is no evidence that China is trying to use Russia as a card to play against the White House. Although China and Russia are making efforts to further promote their partnership, the alliance has not fundamentally changed the strategic pattern.[37] Russia may hope to take advantage of its new relationship with China to shift the balance with the United States in its favor, but this is likely to have little effect as the situation is vastly different from that in the 1970s.

3.2   Unequal Interactive Patterns of Cooperation and Competition in the Triangular Relationship Between China, the United States, and Russia

After the Cold War, with the collapse of the bipolar structure of the United States and the Soviet Union, the triangular relationship also ended and the international structure dominated by major powers commenced

---

[36] Ibid.
[37] Ibid.

a process of transformation and adjustment. The United States, as the winner of the Cold War, is actively trying to build and possess a new world order, the essence of which is to strengthen and pursue its unipolar hegemony. Although Russia is experiencing a difficult domestic political, economic, and diplomatic transition, it seeks to maintain its status as a great power. At the same time, with the remarkable improvement of the global and regional security environment, China has accelerated the pace of its reform and opening up, promoted diplomacy, and participated extensively in the international system. China, the United States, and Russia have gradually formed a new triangular relationship in the game of constructing a new world order of international strategic security and stability.[38]

3.2.1 Dynamic Adjustment of Sino–American–Russian Relations After the Cold War

From a geopolitical perspective, the trend in relations between China, the United States, and Russia after the Cold War shifted in 2011. Until 2011, Russia was the main containment target of the United States. As the only superpower in the post-Cold War era, the United States faced a complicated international situation as a new multipolar pattern formed. Since the collapse of the Soviet Union, Russia has faced a range of security threats emerging from ethnic and religious conflicts, which have even spilled over to peripheral regions of Russia and harmed the interests of the United States and its allies. At the same time, to curb Russia's threat to the United States, Washington has promoted the denuclearization of Ukraine, Belarus, and Kazakhstan by supporting a series of domestic reforms in the fields of political democratization, economic marketization, and the policy of arms reduction. In addition, the United States has greatly promoted the expansion of NATO and the European Union, which squeezes Russia's geostrategic space in Europe. Therefore, Russia was forced to suspend any plan to join Western international organizations and jointly lead European security affairs with the United States.[39] After President Putin took office, he began to concentrate on solving

---

[38] Mastanduno, M. (2019). Partner politics: Russia, China, and the challenge of extending US hegemony after the Cold War. *Security Studies, 28*(3), 479–504.

[39] Goldgeier, J. M., & McFaul, M. (2003). *Power and purpose: US policy toward Russia after the Cold War.* Brookings Institution Press.

Russia's domestic problems and took advantage of the opportunity of the American anti-terrorism focus after 9/11 to form an anti-terrorism partnership with the United States. However, the United States never gave up its plan to contain Russia. Instead, it intervened in Russia's regional political security by planning a series of "color revolutions" and promoted NATO's expansion in Ukraine and Georgia by building anti-missile bases in the Czech Republic and Poland.[40,41] Thus, Russia's strategic security interests have long been seriously threatened by the United States. After the outbreak of the global financial crisis in 2008, the United States and Russia rebooted their relations but there were still major contradictions and fierce conflicts between them on regional security issues.

Unlike the bilateral relations between Russia and the United States, Sino–American relations slightly eased after the end of the Cold War. The long-term policy of the United States toward China is one of both containment and cooperation. The United States seeks to contain China as it does not expect the country to develop so rapidly as to challenge the existing order in the Asia–Pacific region. Meanwhile, Washington hopes to guide Beijing into the international order dominated by the United States instead of having it become a challenger to that order. The common goal of China and the United States of fighting against the Soviet Union having disappeared, conflicts over ideology and the Taiwan issue began to gain importance after the Cold War.[42] As China's strategic interest is mainly focused on domestic economic development, it avoids confrontation with the United States.[43,44] Although there are still frictions between the two sides from time to time, the relations between the two countries are relatively stable as the result of the ever-expanding scale of economic and trade cooperation. However, the financial crisis in 2008

[40] MacMillan, S., & Outlook, N. E. (2015). *Coups and "Color Revolutions": US wages geopolitical warfare against Russia in Central Asia and Caucasus.* Global Research, 9.

[41] Wilson, J. L. (2010). The legacy of the color revolutions for Russian politics and foreign policy. *Problems of Post-Communism, 57*(2), 21–36.

[42] Zhao, Q. (2005). America's response to the rise of China and Sino-US relations. *Asian Journal of Political Science, 13*(2), 1–27.

[43] Gu, X., & Sheng, L. (2010). A sensible policy tool for Pareto improvement: Capital controls. *Journal of World Trade, 44*(3).

[44] Sheng, L., & Zhao, W. (2016). Strategic destination management in the face of foreign competition: The case of Macao SAR. *Journal of Travel & Tourism Marketing, 33*(2), 263–278.

increased friction in Sino–American relations, for two reasons. First, the United States sought China's assistance to purchase its Treasury bonds to ease the crisis and help the United States escape from the economic downturn. Second, the global economic slowdown and the imbalance in capital flows between the United States and China caused by the financial crisis had a major impact on the evolving strategic competition between the two parties.[45]

In the meantime, the bilateral relations between China and Russia have become relatively more friendly since the collapse of the Soviet Union. Driven by the aggressive containment policy of the United States, China and Russia continue to leap over obstacles to establish a strategic partnership that can respond to the unipolar hegemony of America, maintain international strategic stability, and accelerate the formation of a multipolar world system.[46] In 1993, China and Russia began to regard each other as friendly countries; in 1994, they established a constructive partnership; in 1996, the strategic partnership was formalized; and in 2001, the two countries signed a Treaty of Good-Neighborliness and Friendly Cooperation, which established the legal principle that China and Russia will "live in lasting friendship and will never be adversaries against one another."[47] The Kosovo War also alerted China and Russia to the importance and urgency of resisting unipolar hegemony and unilateralism. The Shanghai Cooperation Organization (SCO) and other regional cooperation mechanisms involving the two countries were promoted and established accordingly.[48]

Another significant tipping point for international politics was in 2011, when President Obama announced the strategy of a "Pivot to Asia," which intended to concentrate the United States' military, economic, and diplomatic resources in the Asia–Pacific region to contain China's rise and maintain its leading position in regional security and economic affairs.

---

[45] Friedberg, A. L. (2010). Implications of the financial crisis for the US–China rivalry. *Survival, 52*(4), 31–54.

[46] Hsu, J.-Y., & Soong, J.-J. (2014). Development of China-Russia relations (1949–2011) limits, opportunities, and economic Ties. *Chinese Economy, 47*(3), 70–87.

[47] Turner, S. (2009). Russia, China and a multipolar world order: The danger in the undefined. *Asian Perspective, 33*(1), 159–184.

[48] Hansen, F. S. (2012). China, Russia, and the foreign policy of the SCO. *Connections, 11*(2), 95–102.

China then became the main target of the containment foreign policy of the United States, as Russia took a back seat.[49]

The competition between China and the United States in the areas of geopolitics, economics, and high technology has since intensified day by day. The Pivot to Asia strategy and the accompanying "Asia–Pacific Rebalancing" strategy launched by the Obama administration to some extent strengthened the presence of the US Navy and Air Force while also enhancing its military cooperation with allies in the Asia–Pacific region to consolidate the leading position of the United States in regional security. Washington also encouraged relevant countries to sign the Trans-Pacific Partnership Agreement (TPP), with the goal of formulating new trade rules in the Asia–Pacific region and building a closed regional trade group that would exclude China.

Although President Trump was not keen on the TPP, China was once again identified as a strategic competitor to the United States by the Trump administration. After taking office, Trump released a series of policy documents targeting China, including a new National Security Strategy (NSS) covering politics, economics, military affairs, and diplomacy. The Trump administration implemented the Indo-Pacific Strategy and launched a trade war to further constrain the rise of China.[50] Despite the huge divergences from his predecessor, Joe Biden has continued Donald Trump's anti-China policies to secure the dominance of the United States in the Asia–Pacific region and in the military, political, economic, and high technology domains. In other words, Sino–American trade conflicts appear likely to remain constant for the foreseeable future.

During the same period, the United States and Russia have also moved toward long-term confrontation. Since the Ukrainian crisis, the United States and its European allies have imposed economic sanctions and made military and political containment efforts against Russia, with the economic sanctions against Russia mainly focused on the financial, energy, and military industries having plunged the Russian economy into recession.[51] The bilateral relations between Russia and the United States

---

[49] Logan, J. (2013). *China, America, and the pivot to Asia.* Cato Institute Policy Analysis (717).

[50] Pant, H., & Parpiani, K. (2020). *US engagement in the Indo-Pacific: An assessment of the Trump era.* Observer Research Foundation (ORF).

[51] Mastanduno, M. (2019). Partner Politics: Russia, China, and the Challenge of Extending US Hegemony after the Cold War. *Security Studies, 28*(3), 479–504.

were thus led into an antagonistic state. During the Trump administration, the confrontation between Russia and the United States even escalated into a "hybrid war" across the fields of politics, economics, energy, finance, technology, information, ideology, military security, and international affairs.[52] Joe Biden has continued to implement economic sanctions against Russia, strengthened NATO's military and expanded its armaments in Eastern Europe to suppress Russian troops, and funded the opposition in Commonwealth of Independent States (CIS) countries. The Biden administration's strategy toward Russia may be heading for some slight correction, but the key goal of containment will not change. Against this background, a number of scholars have pessimistically pointed out that it will be difficult for Russia and the United States to normalize their bilateral relations for a long period.[53,54,55]

In contrast to the deterioration of the relations of the United States with both Russia and China, the bilateral relations between China and Russia have developed rapidly over the past 10 years, and their cooperation has been further enhanced in a comprehensive manner in military affairs, energy, economics, and trade. China and Russia have not changed their views on the international order, with both countries opposed to a unipolar structure and committed to jointly safeguarding the international order under the United Nations and international law, which has laid the foundation for their consistent cooperation in international affairs.[56]

---

[52] Trenin, D. (2018). *Avoiding US-Russia Military escalation during the hybrid war*. Carnegie Moscow Centre, 25.

[53] Mastanduno, M. (2019). Partner politics: Russia, China, and the challenge of extending US hegemony after the Cold War. *Security Studies, 28*(3), 479–504.

[54] Rumer, E., Sokolsky, R., & Weiss, A. S. (2017). Trump and Russia: The right way to manage relations. *Foreign Affairs, 96*, 12.

[55] Schoen, D. E., & Kaylan, M. (2014). *The Russia-China axis: The New Cold War and America's crisis of leadership*. Encounter Books.

[56] Lukin, A. (2021). The Russia–China entente and its future. *International Politics, 58*(3), 363–380.

3.3   THE CONFIGURATION OF THE NEW STRATEGIC TRIANGLE IN THE ARCTIC AND ITS DEVELOPMENTAL TRENDS

With its location at the intersection of the three global strategic centers of North America, East Asia, and Europe, the Arctic has been an international arena for world powers for centuries. Recently, climate change has further enhanced the strategic value of the Arctic. Competition between the world powers in the region has intensified in geopolitics and military affairs, and has even extended to extra-regional players like China.

The Arctic now appears as another fulcrum of Sino–American–Russian triangular relations. As noted in previous sections, the United States and Russia have been engaged in military-strategic competition in the Arctic for a long time. During the Cold War, the Arctic was the frontier of military confrontation between the United States and the Soviet Union and had the highest deployment density of intercontinental missiles in the world. After the collapse of the Soviet Union, the strategic space in which Russia can operate has been constantly compressed by the Western allies, which directly led to its aggressive movements in the Ukraine and Crimea. After the Crimean crisis in 2014, Russia became determined to strengthen its military presence in the Arctic to counter containment efforts by the US and the Western alliance.[57]

During the Trump administration, the Arctic became a new arena for the game of the major powers. Although Arctic issues did not enter the government's priority policy agenda, the Trump administration paid increasing attention to these issues. A series of policies and plans were formulated and implemented with the goal of strengthening its military presence in the region and curbing the two major rivals in China and Russia. Since Joe Biden took over in the White House, increasing attention is being paid to the Arctic and Russia. The United States maintains sanctions against Russia, which it identifies as its most direct and urgent threat. In these circumstances, the game between the United States and Russia in the Arctic has become more of a stalemate. With the unprecedented tension between Russia and the United States, the contest between the two countries in the Arctic region has continued to heat up. The Arctic's strategic position and the potential economic benefits from

---

[57] Bertelsen, R. G., & Gallucci, V. (2016). The return of China, post-Cold War Russia, and the Arctic: Changes on land and at sea. *Marine Policy, 72,* 240–245.

waterway development and natural resources are major elements of this competition.[58]

The Bering Strait is the most acutely disputed topic between the United States and Russia because of its role in the opening of the Arctic waterway. The Bering Strait is a narrow sea between the Chukotka Peninsula in Russia and Alaska in the United States. Whether it is the Northwest Passage passing Russia or the Northeast Passage through the Canadian Arctic Islands, any sea route through the Arctic must inevitably pass through the Bering Strait to meet the functional requirements for international navigation.[59] The Bering Strait has thus become a mandatory route for accessing the Arctic waterway. The use of the Bering Strait as a breakthrough route to the Arctic will bring huge economic benefits by more closely connecting the Arctic's hydrocarbon and fishing resources, as well as its metal and timber resources, to the global market. Consequently, controlling the Bering Strait may mean controlling new resource channels. Legally, the Bering Strait is defined in the UNCLOS as a strait used for international navigation. However, there is currently no authoritative organization that clearly defines "straits used for international navigation."[60] The legal status of the Bering Strait thus requires at least a consensus between the United States and Russia, paving the way for a conflict between the two countries in their differing perceptions of claims over the Strait. Although there has been no large-scale and fierce conflict between the United States and Russia in the Bering Strait, frictions have been uninterrupted for a long time. The Strait was especially contested during the Cold War and tensions over who can claim the right to use a possible shipping route running through it may rise again as a result of the proposed Arctic sea route.[61]

Meanwhile, China has also expressed interest in the Arctic. In 2013, China was admitted as an official observer of the Arctic Council. Although

---

[58] Bouffard, T., Greaves, W., Lackenbauer, P. W., & Teeple, N. (2020). *North American Arctic Security Expectations in a New US Administration*. NAADSN Strategic Perspectives, 23.

[59] Berkman, P. A., Vylegzhanin, A. N., & Young, O. R. (2016). Governing the Bering Strait region: Current status, emerging issues and future options. *Ocean Development & International Law, 47*(2), 186–217.

[60] Ibid.

[61] Conley, H. A., & Melino, M. (2019). *The implications of US policy stagnation toward the Arctic region*.

China is neither an Arctic state nor a member of the Arctic Council, in recent years it has attached great importance to the region because of its abundant natural resources and strategic position. According to the UNCLOS, the Arctic high seas are common resources shared by all of mankind, which gives non-Arctic states like China an opportunity to participate in Arctic affairs. In January 2018, Beijing's Arctic policy white paper positioned China as a "near-Arctic state" and expressed its ambition to build the Polar Silk Road with other countries through the development and utilization of the Arctic waterway, and to participate in the infrastructure construction, commercial utilization, and normal operation of the waterway. The document positioned China, as a "responsible power," which intends to take an active part in the study and development of the Arctic, as well as in the governance in the region.[62] The Polar Silk Road is in fact a strategic extension of the BRI in the Arctic. China proposed the "Arctic Silk Road" (hereinafter referred to as ASR), thus including the region into the BRI, the key project of modern China and the way to the "Chinese dream" to come true.[63] Therefore, it is necessary for China to gain the support of Russia, which will play a decisive role in China's presence in the region. The realization of the Polar Silk Road will empower China to cooperate more fully with other countries, release the pressure on China stemming from its huge demand for foreign resources, and reduce its dependence on high-risk traditional maritime sea routes in the South China Sea. More importantly, China's international influence will be broadened through its stronger economic and diplomatic influence in the Arctic region.[64] Understandably, the prospect of such a win–win outcome for China and Russia is not appreciated by the United States.

3.4   Summary

In the 1970s, when the United States and the Soviet Union were competing for global hegemony, the involvement of China began to change the bipolar game of the Cold War and led to the formulation

---

[62] Mariia, K. (2019). China's Arctic policy: Present and future. *The Polar Journal, 9*(1), 94–112.

[63] Ibid.

[64] Tillman, H., Yang, J., & Nielsson, E. T. (2018). The polar silk road: China's new frontier of international cooperation. *China Quarterly of International Strategic Studies, 4*(03), 345–362.

of the theory of the strategic triangle, referring to the relations between China, the Soviet Union, and the United States. Several decades later, the Cold War has ended and the Soviet Union has ceased to exist but the theoretical framework of the strategic triangle is still useful for depicting some of the moves of the major powers in international relations. Given that the ties between China and Russia are increasingly close and that the United States has become more alienated from China and Russia in recent years, the triangular relations between China, Russia, and the United States are somewhat of an extension of the previous strategic triangle. Nonetheless, it is important to point out that the strategic triangle between the three powers today is neither the legacy of the "old" Cold War nor an indicator of a so-called New Cold War.

We argue that recent Arctic affairs epitomize the Sino–Russian–American strategic triangle. China is again playing a role of game changer but is no longer the weakest player in the triangle. With its strong influence on international politics and the global economy, China's presence in the North Pole can change the bipolar dominance of the United States and Russia in the region, while the observation of its activities can enrich the theories of a strategic triangle between China, Russia, and the United States.

References

Abdenur, A. E. (2017). Can the BRICS cooperate in international security? *International Organisations Research Journal, 12*(3), 73–93.

Allison, G. (2018, December 14). *China and Russia: A strategic alliance in the making*. Harvard Kennedy School Belfer Center for Science and International Affair Russia Matters. https://www.russiamatters.org/analysis/china-and-rus sia-strategic-alliance-making

Berkman, P. A., Vylegzhanin, A. N., & Young, O. R. (2016). Governing the Bering Strait region: Current status, emerging issues and future options. *Ocean Development & International Law, 47*(2), 186–217.

Bertelsen, R. G., & Gallucci, V. (2016). The return of China, post-Cold War Russia, and the Arctic: Changes on land and at sea. *Marine Policy, 72*, 240–245.

Blank, S. (2019). Triangularism old and new: China, Russia, and the United States. In J. I. Bekkevold & B. Lo (Eds.), *Sino-Russian relations in the 21st century* (pp. 215–241). Springer.

Bouffard, T., Greaves, W., Lackenbauer, P. W., & Teeple, N. (2020). *North American Arctic Security Expectations in a new US administration*. NAADSN

Strategic Perspectives. https://www.naadsn.ca/wp-content/uploads/2020/11/North-American-Arctic-Security-Expectations-in-a-New-U.S.-Administration-Final.pdf

Carr, E. A., Jr. (2021). *From Trump to Biden and beyond: Reimagining US–China relations*. Springer.

Conley, H. A., & Melino, M. (2019). *The implications of US policy stagnation toward the Arctic Region*. Center for Strategic and International Studies report. https://www.csis.org/analysis/implications-us-policy-stagnation-toward-arctic-region

Cui, S., & Buzan, B. (2016). Great power management in international Society. *The Chinese Journal of International Politics, 9*(2), 181–210.

Friedberg, A. L. (2010). Implications of the financial crisis for the US–China rivalry. *Survival, 52*(4), 31–54.

Gaddis, J. L. (1992). International relations theory and the end of the Cold War. *International Security, 17*(3), 5–58.

Gaddis, J. L. (1997). *We now know: Rethinking cold-war history*. Cambridge University Press.

Goldgeier, J. M., & McFaul, M. (2003). *Power and purpose: US policy toward Russia after the Cold War*. Brookings Institution Press.

Graham, T. (2020). China-Russia-US relations and strategic triangles. *Political Studies, 6*(6), 62–72.

Gu, X., & Sheng, L. (2010). A sensible policy tool for Pareto improvement: Capital controls. *Journal of World Trade, 44*(3).

Hansen, F. S. (2012). China, Russia, and the foreign policy of the SCO. *Connections, 11*(2), 95–102.

Holmes, C. (1979). The Soviet Union and China. In *The Soviet Union in East Asia*. Routledge.

Hsu, J.-Y., & Soong, J.-J. (2014). Development of China-Russia relations (1949–2011) limits, opportunities, and economic ties. *Chinese Economy, 47*(3), 70–87.

Huebert, R. (2019). The New Arctic Strategic Triangle Environment (NASTE). *Breaking the ice curtain? Russia, Canada, and Arctic security in a changing circumpolar world, Canadian Global Affairs Institute report* (pp. 75–93). https://d3n8a8pro7vhmx.cloudfront.net/cdfai/pages/4193/attachments/original/1558816637/Breaking_the_Ice_Curtain.pdf?1558816637

Kissinger, H. (2011). *White house years*. Simon and Schuster.

Kissinger, H. (1969). *National security study memorandum 63*. https://irp.fas.org/offdocs/nssm-nixon/nssm-63.pdf

Logan, J. (2013). *China, America, and the pivot to Asia*. Cato Institute Policy Analysis, 717. https://papers.ssrn.com/sol3/papers.cfm?abstract_id=2228171

Lukin, A. (2021). The Russia-China entente and its future. *International Politics, 58*(3), 363–380.

MacMillan, S. (2015). *Coups and "Color Revolutions": US wages geopolitical warfare against Russia in Central Asia and Caucasus.* Global Research. https://fort-russ.com/2015/04/coups-and-color-revolutions-us-wages/

Mariia, K. (2019). China's Arctic policy: Present and future. *The Polar Journal, 9*(1), 94–112.

Mastanduno, M. (2019). Partner politics: Russia, China, and the challenge of extending US hegemony after the Cold War. *Security Studies, 28*(3), 479–504.

Newlin, C., Conley, H. A., Viakhireva, N., & Timofeev, I. (2020). *US-Russia relations at a crossroad.* Center for Strategic & International Studies (CSIS) report. https://www.csis.org/analysis/us-russia-relations-crossroads

Oxford Analytica. (2021). *Russia-China economic imbalance will deepen.* Emerald Expert Briefings. https://www.emerald.com/insight/content/doi/10.1108/OXAN-DB260047/full/html

Pant, H., & Parpiani, K. (2020). *US engagement in the Indo-Pacific. An assessment of the Trump era* (Observer Research Foundation (ORF) occasional paper 279). https://orfonline.org/wp-content/uploads/2020/10/ORF_OccasionalPaper_279_US-IndoPacific.pdf

Philips, S. E., & Keefer, E. C. (2006). *Foreign relations of the United States, 1969–1976, Volume XVII, China, 1969–1972.* https://history.state.gov/historicaldocuments/frus1969-76v17/d40

Ross, R. S. (2015). *China, the United States, and the Soviet Union: Tripolarity and policy making in the Cold War.* M. E. Sharpe.

Rumer, E., Sokolsky, R., & Weiss, A. S. (2017). *Trump and Russia: The right way to manage relations* (Carnegie Endowment for International Peace report). https://carnegieendowment.org/2017/02/13/trump-and-russia-right-way-to-manage-relations-pub-67995

Schoen, D. E., & Kaylan, M. (2014). *The Russia-China Axis: The New Cold War and America's crisis of leadership.* Encounter Books.

Sheng, L. (2012). Dealing with financial risks of international capital flows: A theoretical framework. *Cambridge Review of International Affairs, 25*(3), 463–474.

Sheng, L. (2014a). Capital controls and international development: A theoretical reconsideration. *Global Policy, 5*(1), 114–120.

Sheng, L. (2014b). Economic structure, cost outsourcing and global imbalances. *Journal of Australian Political Economy, 74*, 81–94.

Sheng, L. (2016). Explaining US-China economic imbalances: A social perspective. *Cambridge Review of International Affairs, 29*(3), 1097–1111.

Sheng, L. (2017). Explaining urban economic governance: The city of Macao. *Cities, 61*, 96–108.

Sheng, L. (2021) *How Covid-19 reshapes new world order: Political economy perspective*. Springer.

Sheng, L., & do Nascimento, D. F. (2021) *The Belt and Road Initiative in South-South Cooperation: The impact on world trade and geopolitics*. Palgrave Macmillan.

Sheng, L., & do Nascimento, D. F. (2021). *Love and trade war: China and the US in historical context*. Springer.

Sheng, L., & Zhao, W. (2016). Strategic destination management in the face of foreign competition: The case of Macao SAR. *Journal of Travel & Tourism Marketing, 33*(2), 263–278.

Statista. (2021). *Value of Russian trade in goods (export, import, and trade balance) with China from 2007 to 2020*. https://www.statista.com/statistics/1003171/russia-value-of-trade-in-goods-with-china/

Sushentsov, A. A., & Suchkov, M. A. (2019). *The nature of the modern crisis in US-Russia relations* (Russia in Global Affairs report). https://eng.globalaffairs.ru/articles/the-nature-of-the-modern-crisis-in-u-s-russia-relations/

Tillman, H., Yang, J., & Nielsson, E. T. (2018). The Polar Silk Road: China's new frontier of international cooperation. *China Quarterly of International Strategic Studies, 4*(03), 345–362.

Trenin, D. (2018). *Avoiding US-Russia military escalation during the hybrid war* (Carnegie Moscow Centre report). https://carnegiemoscow.org/2018/01/25/avoiding-u.s.-russia-military-escalation-during-hybrid-war-pub-75277

Turner, S. (2009). Russia, China and a multipolar world order: The danger in the undefined. *Asian Perspective, 33*(1), 159–184.

Varisco, A. E. (2013). *Towards a multi-polar International System: Which prospects for global peace?* E-International Relations Students, 3.

Wilson, J. L. (2010). The legacy of the color revolutions for Russian politics and foreign policy. *Problems of Post-Communism, 57*(2), 21–36.

Zhao, Q. (2005). America's response to the rise of China and Sino-US relations. *Asian Journal of Political Science, 13*(2), 1–27.

CHAPTER 4

Sino–Russian Cooperation on the Polar Silk Road: Vision, Divergence, and Challenges

Abstract This chapter introduces the nature and history of the Polar Silk Road proposal and explores the vision and challenges of Sino-Russian cooperation on the initiative. Driven by similar pressure from the United States and its allies, China and Russia are strengthening their ties and establishing a series of cooperation projects in the Arctic region. The continuous deepening of Sino-Russian cooperation in the region will provide China with favorable opportunities to become more deeply involved in Arctic affairs.

Keywords Sino–Russian Relations · Polar Silk Road · Marine governance · Belt and Road Initiative

Over the last decade, China has expressed a strong interest in Arctic affairs as it looks to diversify its trade routes and energy supplies and reduce its reliance on traditional routes, such as the Straits of Malacca and the South China Sea. The first sign of this interest came in 2013, when China officially became an observer state of the Arctic Council. In 2015, "the security of China's activities, property, and other interests in the North Pole region" became part of the national strategy according to the State Security Law. In 2018, China released an Arctic policy white paper claiming that it was a "near-Arctic country" and "an

© The Author(s), under exclusive license to Springer Nature Singapore Pte Ltd. 2022
E. L. Sheng, *Arctic Opportunities and Challenges*,
https://doi.org/10.1007/978-981-19-1246-7_4

important shareholder in the Arctic." The policy document also outlined a series of key areas for development and established the relationship between the proposed Polar Silk Road and the BRI. Since the history of China's economic reform indicates that the country never lacks institutional innovation,[1] it will be interesting to observe China's policies in the Arctic.

However, as an extra-regional country, China is "geographically disadvantaged" in Arctic affairs. Its need to gain the support of regional states has driven China to form closer ties with Russia.[2] More specifically, China has expressed a strong interest in cooperating with Russia on developing a northern route. In 2017, China accepted Russia's proposal for it to join in the construction of the Polar Silk Road. Projects included in the Polar Silk Road include developing the Northeast Passage in the Arctic Ocean along the coast of Russia and enhancing the regional connectivity between Russia's Far East and China's northeastern regions. However, there remain a series of divergences between the two countries as the result of different perspectives and contradictory national interests. In this chapter, we analyze three of these divergences. First, Russia's Arctic policy is essentially exclusive, so it is cautious in its cooperation with China in sharing techniques and supporting specific policies. Second, Russia's sovereign claim not only contradicts China's wish to participate in the Arctic as an extra-regional country but also jeopardizes China's image in the Arctic Council. Third, the levels of bilateral trade and economic cooperation have remained relatively low and, given the risky and expensive nature of Arctic development, this constrains the interests of China's investors and constructors. These three risks in Sino–Russian Arctic cooperation should not be ignored.

4.1   China's Arctic Policy and the Polar Silk Road Initiative

China's interest in the Arctic has grown over recent decades, alongside its increasing economic strength and international influence. China has not only been expressing an interest in the Arctic but also taking a more

---

[1] Sheng, L., Li, T., & Wang, J. (2017). Tourism and externalities in an urban context: Theoretical model and empirical evidence. *Cities, 70*, 40–45.

[2] Li, X. P., Fang, Z., & Liu., H. Y. (2017). Rights and obligations of the Arctic Region under the United Nations Convention on the Law of the Sea and their relationship with geographically disadvantaged States. *Polar Research* (2), 279–285.

proactive stance in the region. China used to be a world-class outsourcing destination by offering cost advantages and workforce availability, through which the countries have accumulated a strong economic strength.[3] In 2013, China became a formal observer of the Arctic Council; in the same year, China proposed the Silk Road Economic Belt and the 21st Century Maritime Silk Road, which would later become the BRI. In June 2017, China further expressed its maritime ambitions in its Vision for Maritime Cooperation under the BRI, delivered through the NDRC and SOA.[4]

Although the Vision for Maritime Cooperation does not use the term "Polar Silk Road," it clearly proposes including the joint construction of a "Blue Economic Corridor connecting Europe via the Arctic Ocean" into the BRI, thus setting out a vision for Arctic governance, international scientific cooperation in the investigation, environmental monitoring, improvement, and utilization of waterway conditions, and sustainable development and utilization of resources.[5] The Vision proposes the establishment of three blue economic passages as part of China's BRI maritime cooperation, with the Blue Economic Corridor to pass through the Arctic to Europe and to form a mechanism of geoeconomic construction. Admittedly, the "blue" economic passage through the Arctic Ocean is only an idea put forward by the Chinese government, and it remains unclear where the passage would start or end, but the release of this document was the first time that China had proposed actively promoting the construction of an economic passage connecting it with Europe through the Arctic Ocean.

In October 2017, at the Arctic Circle Assembly in Iceland, which is the largest annual international gathering on the Arctic, Lin Shanqing, Deputy Director of the SOA, pointed out the necessity of strengthening cooperation with Arctic and non-Arctic countries over Arctic affairs. On November 1 of the same year, President Xi Jinping met with visiting Russian Prime Minister Dmitry Medvedev, and China and Russia formally

---

[3] Sheng, L. (2014b). Economic structure, cost outsourcing and global imbalances. *Journal of Australian Political Economy* (74), 81–95.

[4] Yang, J. (2018). The International Environment and Response to the Construction of the Polar Silk Road. *People's Forum· Academic Frontier (Renmin Luntan· Xueshu Qianyan), 2018*(11), 13–23.

[5] Xinhua News Agency. (2017). *Vision for maritime cooperation under the Belt and Road Initiative.*

proposed jointly building the Polar Silk Road.[6] This was the first time that Chinese leaders confirmed in an international forum that China was willing to carry out strategic cooperation with relevant countries in jointly building the Polar Silk Road in the Arctic. The concept of the Polar Silk Road is still developing and changing geographically and in terms of economic cooperation. The vision of the Chinese government is for the Polar Silk Road to form part of the BRI and provide for additional bilateral and multilateral cooperation. Through the BRI and Polar Silk Road, China intends to further cooperate with European and Asian countries to ensure that it has land and sea routes to Europe.

In January 2018, the Chinese government issued an Arctic policy white paper that describes a plan to build the Polar Silk Road in cooperation with various parties around the world. The white paper explains China's identity, responsibilities, principles, and policy propositions in relation to Arctic affairs, with the general theme being that "China is willing to use the Arctic as a link to enhance shared well-being and develop common interests." The white paper expands the geographical area of the Polar Silk Road, which is no longer limited to Sino–Russian cooperation on the building of a Northern sea route and now includes an international shipping channel running through the northern part of Eurasia. This proposed passage would start from the eastern Chinese ports, run north to enter the Arctic Ocean through the Bering Strait, and extend to reach the western ports of Europe, forming a closed loop with the Maritime Silk Road running through the south of Eurasia. From a geoeconomic perspective, the Polar Silk Road is no longer just a project of bilateral cooperation between China and Russia but has become a channel for Arctic multilateral economic cooperation covering East Asia, the Northeast Passage, and countries in Western Europe.

Changes in the global climate and in political and economic conditions, and the further development of cooperation over the Arctic Northeast Passage, could lead to the extension of the Polar Silk Road to the joint development of other Arctic passages. For now, however, the Polar Silk Road is merely a plan for China's future maritime cooperation and geopolitical competition in the Arctic; the details remain unspecified and further policies will be needed to guide its implementation.

[6] *China* Daily. (2017, November 2). Xi backs building of Polar Silk Road.

4.2   The Foundation of Sino–Russian Cooperation on the Polar Silk Road

China and Russia are not only neighboring countries with a long shared border but also close partners with broad areas of common ground. Especially since the Sino–US trade war broke out in 2018, China and Russia have accelerated the pace of cooperation.[7] Arctic affairs have now become another key area of cooperation between China and Russia. As the largest Arctic country, Russia has been the pioneer in exploring the region. Since the days of the Soviet Union, Russia has always assigned a high priority to its Arctic interests. The Soviet Union devised the scenario of a Northern Sea Route (NSR) along the coast of Siberia, running from west of the Kola Peninsula through the Bering Strait in the east (Schøyen & Bråthen, 2011: 978).[8] Russia has insisted on its sovereignty and unilateral control over the waterway for decades and it could be the largest beneficiary of the Polar Silk Road, for two main reasons. First, the Polar Silk Road will enhance transportation links between the East and the West. It will not only benefit Russia as the pathway between the West and the Far East but also shorten the route between Western Europe and East Asia. Second, Russia's Arctic region is rich in oil and gas resources with huge economic value, which could help the development of Russia's Far East. The Russian economy is based on a resource model that is heavily reliant on energy exports, making the long-term stable development of the Russian economy heavily dependent on stable exports of resources and the price expectations of these resources. Tapping the resources in the Arctic region would help Russia to achieve its goal of restoring the country's economic development, expanding its geopolitical influence, and improving its position in the international environment.[9] However, it is impossible for Russia to fully exploit and develop its Arctic territory unilaterally, given a shortage of local labor and advanced technologies, its weak economy, and problems on its periphery related to the Ukrainian

[7] Sheng, L., & do Nascimento, D. F. (2021b). *Love and trade war: China and the US in historical context*. Springer.

[8] Schøyen, H., & Bråthen, S. (2011). The Northern Sea Route versus the Suez Canal: cases from bulk shipping. *Journal of Transport Geography, 19*(4), 977–983.

[9] Komkov, N. I., Selin, V. S., Tsukerman, V. A., & Goryachevskaya, E. S. (2017). Problems and perspectives of innovative development of the industrial system in Russian Arctic regions. *Studies on Russian Economic Development, 28*(1), 31–38.

crisis. The search for international partners such as China is thus a top priority for Russia, especially in the Far East, not only to promote its domestic economic development but also to improve resource extraction technologies and its local labor force.[10]

From China's perspective, although it is not an Arctic country, China's strategic interest in the Arctic has increased alongside its economic development. For decades, China has been heavily reliant on waterways for maritime trade and energy supplies. Domestic social and economic development in China is becoming increasingly dependent on imported resources. China's energy supplies are vulnerable because 80% of its crude oil imports pass through the Straits of Malacca and subsequently the South China Sea.[11] The frequent military activities of the United States in the Asia–Pacific region, its position on the South China Sea, and issues with Taiwan have caused even more disturbances for China.[12] The Island Chain Strategy, proposed by the United States in the 1940s to deter China and the Soviet Union's maritime ambitions,[13] still poses a severe threat to China's maritime trade security, especially in the context of intensifying competition between China and the United States. From Obama's Pivot to Asia to Trump's trade war and Indo-Pacific Strategy, China feels that its southern maritime routes are under threat and thus considers it necessary to diversify its trade routes and energy suppliers. Moreover, China tends to impose strict controls to stem capital outflows and stabilize the RMB, which contradicts America's vision.[14] Therefore, deepening cooperation with Russia in Artic affairs pertaining to energy will help to realize China's energy diversification strategy, improve its ability to cope with energy supply risks, and ensure energy security for the country's further development.

[10] Ellyatt, H. (2019, March 28). *Is Italy playing with fire when it comes to China?* CNBC.

[11] Shambaugh, D. (2018). US-China rivalry in Southeast Asia: power shift or competitive coexistence? *International Security, 42*(4), 85–127.

[12] Sheng, L. (2021). *How Covid-19 reshapes new world order: Political economy perspective.* Springer.

[13] Espena, J., & Bomping, C. (2020). *The Taiwan frontier and the Chinese dominance for the Second Island Chain.* Australian Institute of International Affairs.

[14] Sheng, L. (2014a). Capital controls and international development: A theoretical reconsideration. *Global Policy, 5*(1), 114–120.

With this goal in mind, China has successively opened two corridors—the China–Pakistan Corridor and the China–Myanmar Corridor—to avoid the Straits of Malacca and the South China Sea and connect its western and southwestern territories. Nonetheless, although China's relations with Pakistan and Myanmar have been relatively stable in recent years, there are many uncertain factors within and surrounding these two partners of China.[15] Myanmar and Pakistan both face a series of challenges, such as poverty, insecurity, and deficient infrastructure. In the case of Pakistan, the China–Pakistan Corridor passes through the disputed region of Kashmir, which involves huge security risks and causes concern to India. In the case of Myanmar, conflicts between the government armed forces and local paramilitary groups, frictions among ethnic minorities in the borderlands, and unstable domestic politics are causes of great concern to China. Since early 2021, a military coup, massive protests, and the severity of the COVID-19 pandemic in Myanmar have cast darker shadows over the China–Myanmar Corridor and other infrastructure projects. Meanwhile, the recent blockage of the Suez Canal once again triggered domestic concerns about the safe shipping of energy resources to China. In these circumstances, the continuous pressure of the United States in the South China Sea puts China at risk of being blocked by the First Island Chain—a string of islands, including Taiwan, Okinawa, and the Philippines, that China views as its first line of defense.[16]

During the Trump administration, the United States adopted a much harsher approach to China, including on Arctic issues. The 2018 National Defense Strategy (NDS) states that "Interstate strategic competition is the primary concern in U.S. national security. China is a strategic competitor using predatory economics to intimidate its neighbors while militarizing features in the South China Sea. Russia has violated the borders of nearby nations and pursues veto power over the economic, diplomatic, and security decisions of its neighbors." The United States clearly regards China and Russia as the main threats to its national security. The Biden administration continues to take a hardline approach to China and Russia, as outlined in the Interim National Security Strategic Guidance released in March 2021, which contends that "the distribution of power across the

---

[15] Sheng, L., & do Nascimento, D. F. (2021a). *The Belt and Road Initiative in South-South Cooperation: The impact on world trade and geopolitics.* Palgrave Macmillan.

[16] O'Rourke, R. (2020, March 13). *US-China strategic competition in South and East China Seas: Background and issues for Congress.* Congressional Research Service.

world is changing, creating new threats. China, in particular, has rapidly become more assertive. It is the only competitor potentially capable of combining its economic, diplomatic, military, and technological power to mount a sustained challenge to a stable and open international system. Russia remains determined to enhance its global influence and play a disruptive role on the world stage."

According to a report from *Nikkei Asia*, the United States will "bolster its conventional deterrence against China, establishing a network of precision-strike missiles along this chain as part of $27.4 billion in spending over the next six years."[17] This constitutes the core proposal of the Pacific deterrence initiative submitted to Congress by the United States Indo-Pacific Command. In addition to this military threat by the United States, navigation through the Straits of Malacca is uncertain. China therefore urgently needs to broaden its sea channels to enable it to resist potential external naval operations. For China, the Arctic route will not only shorten sea voyages but also contribute to China's foreign trade and energy security. Therefore, extensive cooperation between China and Russia in the Arctic reflects the concerns of both traditional and non-traditional security. With more stable passages for maritime trade and the transport of energy resources, China is highly motivated to develop Arctic routes via cooperation with Arctic countries such as Russia.

In the current international arena, China and Russia have similar positions and much common ground. The American NSS and NDS identify competition between Russia, China, and the United States as the primary global paradigm. They also depict a world order of zero-sum games and competitive relations. In particular, these strategy documents highlight the politically, strategically, and ideologically confrontational mindset of the United States toward China and Russia. Driven by the disadvantageous external environment, China and Russia, as two powers holding similar positions in the international arena, have been strengthening their bonds and are becoming increasingly close in economic, political, and military affairs as they undertake cooperation in Arctic development.

On May 8, 2015, the Declaration of Cooperation aligning the construction of the Russian-backed Eurasian Economic Union (EAEU) with the Silk Road Economic Belt (SREB) of China was signed in

---

[17] Nakamura, A. (2021). *US to build anti-China missile network along first island chain.* Nikkei Asia.

Moscow. It affirms a deepening of the comprehensive strategic partnership between the two countries as they promote a balanced and harmonious development of the Eurasian region and the world. This partnership includes cooperation on several projects to promote Arctic development.[18] The strong ties between China and Russia will enable them to support each other strategically, and there is no need to worry that their partners may turn to the West, especially during the crisis. This gives Beijing and Moscow more leeway to consolidate their respective regional spheres of influence. On July 2, 2021, President Putin signed a new 44-page Russian national security strategy. This strategy provides hints and presuppositions for Russia's diplomacy and officially marks the United States and its NATO allies as unfriendly countries. Russia's relations with European countries are no longer prioritized, and these countries rank very low in intimacy. The report names China and India as Russia's strategic partners, and expectations are placed on institutions such as the SCO and BRICS, and on Russia–India–China links. Trenin described this strategic statement as "a manifesto for a different era: one defined by the increasingly intense confrontation with the United States and its allies."[19] The improvement of Sino–Russian relations creates favorable external conditions for the cooperation between the two countries in the Arctic region.

The Yamal liquefied natural gas (LNG) project can be viewed as a model of Sino–Russian cooperation in the Arctic. A key energy cooperation project, Yamal LNG, is the world's first integrated project of exploration, development, liquefaction, transportation, and sale of polar natural gas. After the project is completed and put into operation, it is estimated that it will provide approximately 6 billion cubic meters of natural gas to China every year.[20] Russia is expected to utilize overseas funds and international technical cooperation to develop Arctic resources, build the Arctic into a new modern energy base, maintain Russia's position as a major energy exporter, and provide support for Russia's

---

[18] Rozman, G., & Radchenko, S. (Eds.). (2017). *International relations and Asia's northern tier: Sino-Russia relations, North Korea, and Mongolia.* Springer.

[19] Trenin, D. (2021). *Russia's national security strategy: A manifesto for a new era.*

[20] Tillman, H., Yang, J., & Nielsson, E. T. (2018). The polar silk road: China's new frontier of international cooperation. *China Quarterly of International Strategic Studies, 4*(03), 345–362.

economic development (Byers, 2017).[21] China's capital and infrastructure construction technology greatly appeal to Russia and other Arctic countries.

4.3   Uncertainties in Sino–Russian Cooperation: Divergences, Challenges, and Risks

China and Russia have extensive shared interests in many fields. Nonetheless, there are several areas of uncertainty and even confrontation in their bilateral relationship that could pose challenges in Sino–Russian Arctic cooperation. These challenges to Arctic cooperation can be categorized into two types: First, "the yoke of history" between China and Russia could hinder bilateral trust; second, strategic differences between China and Russia could reduce the depth of their cooperation.

4.3.1   "The Yoke of History"

China and Russia are two adjacent great powers, and centuries of proximity have made for a complicated history of bilateral relations. The complexities mainly relate to the territory of the Far East and the history of the Sino–Soviet split. From the nineteenth to the early twentieth centuries, Tsarist Russia coerced the Qing government of China into signing a range of unequal treaties that ceded more than 1.5 million square kilometers of China. This history is humiliating for many Chinese people and has shaped the victim identity of the Chinese in modern history. More than a century later, China and Russia have settled their border disputes and China does not make any claims on its former territory, but the historical problem still upsets the relations between the two countries. In China, many people remain dissatisfied over the loss of territory in the north; Russia, meanwhile, is wary of any Chinese movements in its Far East.

The Sino–Soviet split of the mid-twentieth century is a cautionary tale for Sino–Russian cooperation. Since the Soviet era, Russia's attitude

---

[21] Byers, M. (2017). Crises and international cooperation: An Arctic case study. *International Relations, 31*(4), 375–402.

toward cooperation with China has always been conditional.[22] According to the archives and the *Selected Works of Mao Zedong on Diplomacy*,[23] from the late 1950s to the early 1960s, the relationship between China and the Soviet Union began to deteriorate with a series of demands made by the Soviet Union, such as establishing a longwave radio station in Hainan and a Sino–Soviet joint fleet. China thought that acceding to these demands would put the country at a politically disadvantaged position. At that time, China was already dissatisfied with the aggressive Soviet attitude. China's leader complained that the Soviet Union never had faith in Chinese citizens and treated them as if they were its sons. Moreover, China and the Soviet Union failed to reach an agreement on the border issue. The Soviet Union denied the unequal nature of the treaty and claimed that "historical formation" and "actual guarding" were also the basis for resolving border issues. These principles required China to recognize that the Chinese territories occupied by Tsarist Russia and the Soviet Union in violation of their contract were all Soviet territories. In the same year, many Soviet aid experts withdrew from China, which caused great damage to China's economic development.[24] As the relationship between the two countries deteriorated in the mid-1950s, the Soviet Union increased its troop numbers in the border areas and sent troops to Mongolia, which directly threatened China's security. In 1969, the Chen Pao Island and Tielekti incidents occurred successively, and relations between the two sides evolved from a cold ideological war to a hot border conflict. Although China and Russia have settled their border disputes and established a strategic partnership in the twenty-first century, the ghost of history will haunt the establishment of mutual trust and deep cooperation between the two sides.

---

[22] Stronski, P., & Ng, N. (2018). *Cooperation and competition: Russia and China in Central Asia, the Russian Far East, and the Arctic* (Vol. 28). Carnegie Endowment for International Peace.

[23] "Minutes of Conversation, Mao Zedong and Ambassador Yudin," July 22, 1958, History and Public Policy Program Digital Archive; Zedong, M. (1994). *waijiao wenxuan* (Selected Works of Mao Zedong on Diplomacy), Zhongyang wenxian chubanshe, 322–333, translated and annotated by Zhang Shu Guang and Chen Jian.

[24] Zhang, X. (2019). Facilities connectivity in eastern regions of China and Russia and the "Belt and Road" initiative. *R-economy, 5*(3), 144–150.

4.3.2   Strategic Differences

Sovereignty of the Arctic still hangs in the balance because of a range of intricate controversies. The Arctic countries and the wider international community are in dispute over rights to territory and navigation in the Arctic region, and there is no international treaty uniformly applicable to all parties. Against this background, Russia has been strongly persistent in claiming its sovereign and unilateral control over the NSR as its national transport and communication route subject to national legislation on historical grounds[25] and argues that the ice-covered straits of its Arctic archipelagos are internal waters with the support of several theories, including that of historic bays closed by straight baselines. In Russian legislation, its internal waters also include the waters of the White Sea, the bays of Cheshkaya and Baridaratskaya, and the waters separating the mainland from the archipelagos. Arctic countries, especially America, have engaged in a long-standing dispute over the legal status of the NSR.

In 2001, Russia submitted a 200 nautical mile continental shelf case with the CLCS, which aims to facilitate the implementation of the UNCLOS.[26] The claimed outer continental shelf area is approximately 120 square kilometers, accounting for nearly half of its 200 nautical miles from the Arctic Ocean. However, Russia's claim was rejected due to insufficient evidence. Over the subsequent decade, it submitted repeated applications claiming sovereignty of the Arctic. This action on the part of Russia is regarded as aggressive and has aroused strong dissatisfaction among Arctic countries. It is a fact, nonetheless, that Russia has used its geographic advantage to take control of the northern waterway since the Soviet era. Russia has strongly rejected non-Arctic countries intervening in Arctic affairs, proposing the Nuuk criteria to create obstacles to the participation of extra-regional countries. With its goal of maximizing control of the Arctic, Russia certainly does not want China to challenge its advantageous position in the NSR or in Arctic affairs in general.

Furthermore, Russia's claim on the Arctic waterway contradicts China's vision of the Arctic. For China, all countries are entitled to enjoy

[25] Todorov, A. A. (2017). The Russia-USA legal dispute over the straits of the Northern Sea Route and similar case of the Northwest Passage. *Arctic and North, 29*, 62–75.

[26] Commission on the Limits of the Continental Shelf (CLCS). Outer limits of the continental shelf beyond 200 nautical miles from the baselines: Submissions to the Commission: Submission by the Russian Federation.

the right and freedom of navigation according to international law.[27] Moreover, with the fundamental principle of noninterference, China is cautious about becoming involved in any territorial disputes among Arctic countries. Cooperation with Russia in the Arctic and the NSR could have two severe side effects for China. First, China's relations with other Arctic countries could deteriorate, when considering Russia's aggressive diplomacy over the region. Second, China's increasing presence in the Arctic Ocean might have an impact on the situation in the South China Sea.

An imbalance between supply and demand is another problem for Sino–Russian Arctic cooperation. Chinese investors are reluctant to invest in high-cost and high-risk projects unless they are guaranteed a management role, voice, and vote. However, Russia is unwilling to offer the necessary degree of cooperation. There is a similar issue already existing in relation to aviation rights for connecting Asia and Europe by flying over Russia. As these flights currently must pass over Russia, Russia extracts huge fees for the rights. Russia can be expected to do the same for granting ships the right of passage through "their waterway" in the NSR. Since the publication of China's Arctic policy white paper, the Polar Silk Road initiative has been widely seen in Russia as an opportunity to attract investment for Arctic transportation infrastructure and port facilities and to consolidate Russia's sovereignty over the Arctic. However, there are also public concerns that China intends to control Arctic resources or occupy a dominant position in the Arctic. It has been even pointed out that if China seeks to control Arctic resources, this will hurt the economic interests of Russia and the United States in the region. Predictably, Russia is not satisfied with China's aspirations to manage the Arctic, and Moscow is firmly intent on retaining the privileges granted to the five countries bordering the Arctic Ocean.[28]

Finally, Russia's domestic skepticism toward China's motives in the construction of the Polar Silk Road remains. Russia regards the Arctic region as an important factor to support the country's recovery and considers its future economic development as largely dependent on the economic wealth brought on by the melting of the Arctic glaciers. Russia is therefore wary of other countries being involved in Arctic affairs and

---

[27] The State Council Information Office of the People's Republic of China. (2018). *China's White Paper on Arctic Policy*.

[28] Trenin, D. (2020). *Russia and China in the Arctic: Cooperation, competition, and consequences*. Carnegie Moscow Center.

fears that China will use the development of the Arctic shipping routes to seize its own rights in the Arctic. Accordingly, Russia takes precautions against China while making use of Chinese capital, technology, and markets. Some Russian scholars who believe in the Chinese threat theory fear that Russia's vast energy resources will be invested in China and will become an important energy base for developing and strengthening Chinese industries. In the end, Russia may become a tool for China to counter the United States, leaving Russia to serve the needs of China's economic development and, with dependence on Chinese capital, technology, and markets, to become a "vassal of China." Russia has introduced a variety of plans to deal with this possibility. Specifically, Russia has taken advantage of its dominant position in Arctic affairs to assign the more profitable offshore projects to Russian state-owned enterprises, with foreign companies facing higher barriers to joining these projects. Moreover, in projects that are open to foreign companies and capital, Russia tends to take a multiparty approach to attracting investment. Such plans have adverse effects on the depth of cooperation between China and Russia.

4.4   SUMMARY

Driven by the same pressures from the United States and its allies, China and Russia are strengthening their ties and establishing a series of cooperation projects in the Arctic region. Against this background, the Polar Silk Road has been envisioned to allow for greater integration and complement China's BRI economic corridors. With the continuous deepening of Sino–Russian cooperation, China will continue to have favorable opportunities to become more deeply involved in Arctic affairs. As activities in the Arctic between the two countries intensify, voices highlighting China's Arctic ambitions will be further amplified in the complex and tense international situation.

However, there are still a range of uncertainties in relation to bilateral divergences in Arctic cooperation between China and Russia. First, as a regional outsider, China is reliant on Russia for its presence in the Arctic, but China's positions on the legal status of the Northeast Passage and the continental shelf in the Arctic Ocean are fundamentally divergent from Russia's claims. Considering the aggressive diplomacy of Russia over Arctic issues, the interdependence of China and Russia in the Arctic may lead to the deterioration of relations between China and the other

Arctic countries and may have a negative impact on China's claims in the South China Sea. Second, Russia's domestic politics and conflicts between interest groups bring more uncertainties to Sino–Russian Arctic development. Nonetheless, the divergences that threaten cooperation between China and Russia in the Arctic region do not override the fact that China and Russia are expanding and deepening their cooperation in the Arctic. The frictions have not hindered the development of cooperation between the two countries thus far. Sino–Russian cooperation in the Arctic region is still undergoing dynamic adjustments but there are enormous prospects.

REFERENCES

Byers, M. (2017). Crises and international cooperation: An Arctic case study. *International relations, 31*(4), 375–402.

China Daily. (2017, November 2). *Xi backs building of Polar Silk Road.* https://www.chinadaily.com.cn/china/xismoments/2017-11/02/con tent_34012484.htm

Ellyatt, H. (2019, March 28). *Is Italy playing with fire when it comes to China?* CNBC. https://www.cnbc.com/2019/03/27/italys-joins-chinas-belt-and-road-initiative.html

Espena, J., & Bomping, C. (2020). *The Taiwan Frontier and the Chinese dominance for the Second Island Chain.* Australian Institute of International Affairs, Australian Institute of International Affairs report. https://www.internationalaffairs.org.au/australianoutlook/taiwan-fro ntier-chinese-dominance-for-second-island-chain/

Komkov, N. I., Selin, V. S., Tsukerman, V. A., & Goryachevskaya, E. S. (2017). Problems and perspectives of innovative development of the industrial system in Russian Arctic regions. *Studies on Russian Economic Development, 28*(1), 31–38.

Li, X. P., Fang, Z., & Liu, H. Y. (2017). Rights and obligations of the Arctic Region under the United Nations Convention on the Law of the Sea and their relationship with geographically disadvantaged States. *Polar Research, 2,* 279–285.

Nakamura, R. (2021). *US to build anti-China missile network along first island chain.* Nikkei Asia. https://asia.nikkei.com/Politics/International-relations/Indo-Pacific/US-to-build-anti-China-missile-network-along-first-island-chain

O'Rourke, R. (2020). *US-China strategic competition in South and East China Seas: Background and issues for Congress.* Congressional Research Service R42784. https://crsreports.congress.gov/product/pdf/R/R42784

Rozman, G., & Radchenko, S. (Eds.). (2017). *International Relations and Asia's Northern Tier: Sino-Russia Relations, North Korea, and Mongolia.* Springer.

Schøyen, H., & Bråthen, S. (2011). The Northern Sea Route versus the Suez Canal: Cases from bulk shipping. *Journal of Transport Geography, 19*(4), 977–983.

Shambaugh, D. (2018). US-China rivalry in Southeast Asia: Power shift or competitive coexistence? *International Security, 42*(4), 85–127.

Sheng, L. (2014a). Capital controls and international development: A theoretical reconsideration. *Global Policy, 5*(1), 114–120.

Sheng, L. (2014b). Economic structure, cost outsourcing and global imbalances. *Journal of Australian Political Economy, 74*, 81–95.

Sheng, L. (2015). Theorizing income inequality in the face of financial globalization. *The Social Science Journal, 52*(3), 415–424.

Sheng, L. (2021). *How Covid-19 reshapes new world order: Political economy perspective.* Springer

Sheng, L., & do Nascimento, D. (2021a). *The Belt and Road Initiative in South-South Cooperation: The impact on world trade and geopolitics.* Palgrave Macmillan.

Sheng, L., & do Nascimento, D. F. (2021b). *Love and trade war: China and the US in historical context.* Springer.

Sheng, L., Li, T., & Wang, J. (2017). Tourism and externalities in an urban context: Theoretical model and empirical evidence. *Cities, 70*, 40–45.

Sheng, L., & Zhao, W. (2016). Strategic destination management in the face of foreign competition: The case of Macao SAR. *Journal of Travel & Tourism Marketing, 33*(2), 263–278.

Stronski, P., & Ng, N. (2018). *Cooperation and competition: Russia and China in Central Asia, the Russian Far East, and the Arctic.* Carnegie Endowment for International Peace report. https://carnegieendowment.org/2018/02/28/cooperation-and-competition-russia-and-china-in-central-asia-russian-far-east-and-arctic-pub-75673

The State Council Information Office of the People's Republic of China. (2018). *China's White Paper on Arctic Policy.* http://www.scio.gov.cn/zfbps/32832/Document/1618203/1618203.htm

Tillman, H., Yang, J., & Nielsson, E. T. (2018). The polar silk road: China's new frontier of international cooperation. *China Quarterly of International Strategic Studies, 4*(3), 345–362.

Todorov, A. A. (2017). The Russia-USA legal dispute over the straits of the Northern Sea Route and similar case of the Northwest Passage. *Arctic and North, 29*, 62–75.

Trenin, D. (2020). *Russia and China in the Arctic: Cooperation, competition, and consequences.* Carnegie Moscow Center report. https://carnegiemoscow.org/commentary/81407

Trenin, D. (2021). *Russia's national security strategy: A manifesto for a new era.* Carnegie Moscow Center report. https://carnegiemoscow.org/commentary/84893

Xinhua News Agency. (2017). *Full Vision for maritime cooperation under the Belt and Road Initiative.* http://www.xinhuanet.com/english/2017-06/20/c_136380414.htm

Yang, J. (2018). The international environment and response to the construction of the Polar Silk Road. *People's Forum· Academic Frontier (Renmin Luntan· Xueshu Qianyan), 2018*(11), 13–23.

Zhang, X. (2019). Facilities connectivity in eastern regions of China and Russia and the "Belt and Road" initiative. *R-economy, 5*(3), 144–150.

CHAPTER 5

Participation in and Strategic Arrangements for Arctic Affairs by the United States: Seeking Collective Security in a New Era

Abstract As the key player in the game surrounding the North Pole, the United States has made a series of adjustments to its Arctic policy from the Second World War and through the Cold War to the present. The strategic significance of the Arctic to the United States is currently increasing amidst simultaneous tensions with China and Russia. The three countries, which form a New Arctic Strategic Triangle and are now experiencing the coldest period of their relations since the end of the Cold War, will continue to play a fierce game in the freezing Arctic Ocean.

Keywords Arctic strategy · Sino–US competition · Polar Silk Road · US–Russian relations

The two neighboring powers of the United States and Russia eye each other across the Arctic Ocean. During the Cold War, the Arctic was highly militarized as the frontline of confrontation between the two superpowers. However, the Arctic policy of the United States has been relatively inconsistent, especially in the post-Cold War era. With its Arctic territory of Alaska located some distance from the other states, the Arctic tends to be "out of sight, out of mind" for policymakers in the United States. Furthermore, its sound geographic conditions mean that the United States is not as keen as Russia to exploit the Arctic region.

© The Author(s), under exclusive license to Springer Nature Singapore Pte Ltd. 2022
E. L. Sheng, *Arctic Opportunities and Challenges*,
https://doi.org/10.1007/978-981-19-1246-7_5

However, driven by a series of changes in its international relations, the United States has recently adjusted its Arctic policy. Given its dual role as a global power and one of the eight Arctic countries, the United States is empowered to determine the Arctic strategic decision-making process, but for successive American governments, this has been limited to the framework of its global strategy. For a long time, the United States has focused its global strategy mostly on the Atlantic and the Pacific, with little attention to the Arctic region and relatively limited strategic investment in this area. The limitations of its strategic vision mean that the United States has been unable to occupy a dominant position in the game of Arctic geopolitics and geoeconomics.

5.1   The Development and Adjustment of the American Arctic Strategy

During the Trump administration from 2017 to 2021, the United States made a series of adjustments to its Arctic policy and reaffirmed the significance of the Arctic: "A range of international institutions establishes the rules for how states, businesses, and individuals interact with each other, across land and sea, the Arctic, outer space, and the digital realm. It is vital to U.S. prosperity and security that these institutions uphold the rules that help keep these common domains open and free." In 2019, the Commander of the United States Air Force upgraded the Arctic as a matter of substance: "The Arctic now is America's first line of defense."[1]

The competition between different economic and political factors in developed countries and emerging countries in a globalized world often leads to asymmetric or unfair bargaining outcomes as a result of strategic interactions.[2,3,4] Most scholars from America think tanks have examined the Polar Silk Road initiative with a competitive mindset and argued that

---

[1] Sheng, L., & do Nascimento, D. F. (2021b). *Love and trade war: China and US in historical context*. Springer.

[2] Sheng, L. (2014a). Capital controls and international development: A theoretical reconsideration. *Global Policy, 5*(1), 114–120.

[3] Sheng, L., & Zhao, W. (2016). Strategic destination management in the face of foreign competition: The case of Macao SAR. *Journal of Travel & Tourism Marketing, 33*(2), 263–278.

[4] Sheng, L., Li, T., & Wang, J. (2017). Tourism and externalities in an urban context: Theoretical model and empirical evidence. *Cities, 70*, 40–45.

China's proposal has geostrategic motives. Obviously alarmed by the close cooperation between China and Russia in the Arctic, the United States is securitizing Arctic affairs and is strengthening its military presence in the region with frequent army drills. Its tensions with China and Russia increase the strategic significance of the Arctic for the United States. It is believed that Alaska will be the bridgehead to deter China and Russia. With the change of administration of the United States, it seems that Arctic issues are one of the few common interests between Joe Biden and Donald Trump and that fierce competition will therefore continue. The country is also plagued by thorny domestic issues such as social inequality and excessive financial liberalization, which has led some politicians to shift domestic attention to international affairs.[5] As a global power, the strategic goals of the United States are based on its global interests. China's BRI, even if it is not designed on geopolitical considerations, undoubtedly has geopolitical and geoeconomic effects.[6] Conflicts with the interests of the United States are therefore inevitable.

Toward the end of the Cold War, Gorbachev's reforms made cooperation possible in the Arctic.[7] Canada first proposed to set up an Arctic Council in 1989, intending the body to cover a broad range of topics, but received negative feedback from the Reagan administration, which preferred a unilateral or bilateral approach to the region. However, the cold response from the United States did not discourage Canada, and it later proposed to create the Arctic Environmental Protection Strategy, to be mainly focused on environmental issues, with cooperation from Finland. In 1995, the United States eventually expressed its support for the Arctic Council under the Clinton administration, but strongly insisted that attention should be paid only to environmental issues. Thus, the footnotes of the final agreement specifically state that the Arctic Council should not deal with military security issues. The willingness of the United

---

[5] Yin, Y. C., & Sheng, L. (2021). Theorizing about global imbalances: An inequality perspective. *Argumenta Oeconomica*, 46, 169–181.

[6] Sheng, L., & do Nascimento, D. F. (2021b). *Love and trade war: China and the US in historical context*. Springer.

[7] Huebert, R. (2009). *United States Arctic policy: The reluctant Arctic power*. University of Calgary, The School of Public Policy–University of Calgary Publications Series, 2(2).

States to cooperate on environmental protection was driven by rising environmental awareness.[8] In March 1989, before discussions with the Arctic states in Rovaniemi, there was the disastrous Exxon Valdez oil spill in Alaska. At the same time, there was a rising trend of climate change awareness following Professor James Hansen's testimony in 1988. With the end of the Cold War, concerns about climate change and the vulnerability of the Arctic environment were able to shape the Arctic strategy set out by the United States in 1994. The policy emphasized the importance of maintaining defense capabilities and marine access in the Arctic to fulfill post-Cold War security needs, but also emphasized the need for environmental protection and to balance resource management and economic development.

On January 9, 2009, before the Bush administration left office, the United States released another policy regarding the Arctic. It was a landmark in the Arctic policy of the United States because it divided the polar policy of the United States into the Arctic and the Antarctic.[9] Like the 1994 Arctic policy, the new policy listed six primary objectives, which were largely identical to the old goals except that homeland security was added to national security and the objective to meet security needs was moved from the bottom of the list to the top. The emphasis on homeland security, made in the context of the 9/11 terrorist attack, signaled a shift in America's perception of the Arctic. "Domain awareness," which had emerged as a new term under the homeland security concept, brought attention to American interests in the Arctic amid environmental changes.[10]

During the Obama administration, the United States strengthened its focus on the Arctic because of the issue of climate change, which was an important part of Obama's presidential campaign. While he was in office, Barack Obama made great efforts to strike a balance between environmental concerns and economic opportunities in the Arctic. Obama's preference for multilateralism made the United States more engaged in circumpolar cooperation, as stressed in the 2013 National Strategy for

---

[8] Haycox, S. (2020). Arctic policy of the United States: An historical survey. In *The Palgrave handbook of Arctic policy and politics* (pp. 233–250). Palgrave Macmillan.

[9] Huebert, R. (2009). *United States arctic policy: The reluctant arctic power*. University of Calgary, The School of Public Policy–University of Calgary Publications Series, 2(2).

[10] Nilsson, A. E. (2018). The United States and the making of an Arctic nation. *Polar Record, 54*(2), 95–107.

the Arctic Region. To ensure the sustainable development of the region, the United States actively worked in partnership with other Arctic states, with the Obama–Trudeau agreement with Canada as an example.

However, most of Obama's efforts were undone by Donald Trump.[11] The denial of climate change was one of the dramatic aspects of Trump's presidential campaign, which Trump put into action after entering the White House by undoing many of Obama's environmental policy initiatives, including those concerning the Arctic.[12] Under the Trump administration, the United States shifted the focus of its Arctic policy from environmental protection to exploiting economic potential as it sought to fulfill Trump's campaign commitment to creating jobs for Americans.

Donald Trump not only downsized the Environmental Protection Agency and reduced international funding for climate research and sustainable development but also permitted offshore hydrocarbon production in the Arctic and leased the country's Arctic waters for offshore oil and gas extraction. Despite refusing to use the term "climate change" in strategic documents, the consequences of climate change continued to have an influence on the United States' Arctic policy under the Trump administration.[13] For example, the Strategic Outlook for the Arctic published by the Navy in January 2019 began by addressing the increase in maritime traffic and resource accessibility due to environmental changes in the Arctic. It concluded that the Navy needed to be prepared to protect national interests in the region and placed an emphasis on strategic competition. Its increased military presence in the Arctic was also driven by the "great power competition" perceived by the United States. The Arctic strategy released by the Department of Defense in the same year clearly named Russia and China as the main competitors to the United States, both globally and regionally in the Arctic. In the Arctic, it stressed the importance of achieving competitive military advantages and maintaining credible deterrence. To a large extent, Trump's Arctic policy was a concrete reflection of the guiding ideology of his national security

[11] Sheng, L. (2021). *How Covid-19 reshapes new world order: Political economy perspective*. Springer.

[12] Käpylä, J., & Mikkola, H. (2019). Contemporary Arctic meets world politics: Rethinking Arctic exceptionalism in the age of uncertainty. In *The Global Arctic Handbook* (pp. 153–169). Springer.

[13] Lavorio, A. (2021). Geography, climate change, national security: The case of the evolving US Arctic strategy. *The International Spectator, 56*(1), 111–125.

strategy in the Arctic region: that the United States should try its best to maintain its leadership and rule-making power in Arctic affairs. Guided by such a strategy, it would have been unthinkable for the Trump administration to let competitors such as China and Russia dominate Arctic affairs and governance.

The Navy's 2019 Strategic Outlook for the Arctic details the recognized threats, opportunities, and risks facing the United States in the Arctic region and demonstrates the ability to peacefully resolve differences and conflicts with all nations. It states that the Navy will act to safeguard the country's Arctic interests and promote a strategy for Arctic security.[14] In June of the same year, the Office of the Under Secretary of Defense for Policy of the United States released a new edition of the Department of Defense Arctic Strategy, outlining its methods for protecting the national security and interests of the United States in the Arctic region in the new era of strategic competition. Compared with the previous Arctic Strategy released by the Ministry of National Defense in 2013, the focus of the new report has shifted from protecting the Arctic environment to strengthening Arctic security. This strategy focuses on competition with China and Russia, holding that these two countries are the main challengers to the long-term security and prosperity of the United States.[15] Given that both Russia and China have identified the Arctic region as a national priority and have invested significant resources to expand their building capacity and influence in the Arctic, the United States believes that the challenges posed by Russia and China to the global order will inevitably also affect the order of peace in the Arctic. The report also points out that its allies and partners share common national interests in the Arctic order and that this is the greatest strategic advantage of the United States in the Arctic region. However, although cooperation between the United States and its Arctic allies and partners is said to strengthen cooperation on regional security, it thereby encourages strategic competitors to attempt to unilaterally change the rules-based order.[16]

---

[14] United States Navy. (2019). *Strategic outlook for Arctic*, 14.

[15] Sheng, L., & do Nascimento D. F. (2021a). *The Belt and road initiative in South-South cooperation: The impact on world trade and geopolitics*. Palgrave MacMillan.

[16] Ibid.

The United States is also becoming more vigilant about the participation of China and Russia in Arctic affairs. The Department of Defense believes that although China has no territorial claims in this region, it is seeking a certain influence in Arctic governance. As part of China's BRI, it has linked its economic activities in the Arctic to its broader strategic objectives, as stated in 2018 in its first white paper on Arctic policy.[17] The white paper indicates that China's interest in the Arctic is mainly to obtain natural resources and exploits the opportunities provided by Arctic sea routes for shipping and investing in strategic sectors and scientific activities in Arctic countries. China has research stations in Iceland and Norway and carries out energy development and infrastructure projects in Russia, such as the Yamal LNG project. It is believed that China will continue to seek opportunities to invest in dual-use infrastructure in the Arctic. At the same time, the U.S. Department of Defense is concerned about the Arctic participation of Russia. Since the establishment of the Northern Fleet Joint Strategic Command in Russia in 2014, Moscow has strengthened its presence in the Arctic by renovating its airport, building new military bases, and building an air defense system. In addition, Russia's commercial investment and sustained defense investment in the Arctic region have strengthened its territorial defense and ability to control the northern sea lanes. China and Russia pose different challenges in their respective fields, but the Arctic activities and capabilities pursued by the two countries may bring risks to the United States.[18] Therefore, it is necessary to adjust the strategic focus of the United States in relation to the Arctic to counter the growing influence of China and Russia in the region.[19]

In addition to military and strategic changes, there has been a shift in the resource development strategy of the United States in the Arctic region. Before Trump took office, the exploitation of American Arctic oil and gas resources was relatively low and limited to the jurisdiction of Alaska, with federal waters having not yet been developed. For the purpose of environmental protection, the Obama administration

[17] Mariia, K. (2019). China's Arctic policy: Present and future. *The Polar Journal, 9*(1), 94–112.

[18] Allison, G. (2018, December 14). *China and Russia: A strategic alliance in the making*. The Center for the National Interest.

[19] McCann, K. B., McMullin, J. A., & Turvold, W. D. (2020). *Before it's too late: US Maritime Grand Strategy in the Indian Ocean and the Arctic Ocean*. Daniel K. Inouye Asia–Pacific Center for Security Studies.

announced a permanent ban on the development of offshore oil and gas resources in American Arctic waters, such as the Chukchi and Beaufort Seas.[20] Five days after Trump took office as president, however, he issued an executive order for the America First Energy Plan. The order announced that the United States needed to not only achieve energy independence but also place energy in the leading position as the strategic goal of its economic and foreign policy. The Ministry of the Interior was asked to reassess the ban that was placed on drilling in Arctic Ocean waters during the Obama administration and promote the opening of oil exploration in the Arctic region.[21] On October 24, 2018, the Trump administration approved the first oil exploitation project in the Arctic federal waters and agreed to the development of oil extraction facilities in the Beaufort Sea and the east of Prudhoe Bay by the Texas-based Hilcorp energy company. Trump's strong policy ushered in new opportunities for American Arctic oil and gas development.[22]

In 2017, Trump also announced the withdrawal of the United States from the Paris Agreement, which was signed by more than 200 countries and aimed to control the rise in global average temperatures in the twenty-first century to within 2 degrees Celsius compared with the pre-industrial level and to strive to control the rise to within 1.5 degrees Celsius.[23] On November 4, 2020, the United States became the first and only country to formally withdraw from the agreement, sparking great discontent within the international community.[24] This further removed obstacles for energy development in the Arctic region of the United States. On the basis that human activity is not a major driver of current global warming, the Trump administration made significant budget cuts in areas related to climate policy and research and rescinded four previous

[20] Conley, H. A., & Melino, M. (2019). *The implications of US policy stagnation toward the Arctic region*. JSTOR.

[21] Vakhshouri, S. (2017). *America first energy plan: Renewing the confidence of American energy producers*.

[22] Harball, E. (2018, October). *Trump administration approves first oil production in federal Arctic waters*.

[23] Urpelainen, J., & Van de Graaf, T. (2018). United States non-cooperation and the Paris agreement. *Climate Policy, 18*(7), 839–851.

[24] Arlota, C. (2021). *Cost & benefit analysis of the United States' withdrawal from the Paris Agreement*.

presidential executive orders related to climate change.[25] Since President Biden took office, there has been some changes in American environmental policy. For example, in the first half of 2021, he announced that he would stop oil development in the Arctic National Wildlife Refuge in Alaska, reversed the decision of former President Trump to allow drilling in the Arctic, and rejoined the Paris Agreement.[26] Additionally, Biden has proposed plans for a clean energy revolution, with a total investment of up to US$400 billion over 10 years to drive a transition to a 100% clean energy economy with net zero emissions by 2050. This will constrain the activities of the traditional energy sectors, such as the oil and gas industries, that the Trump administration had focused on in the Arctic region.[27] The future development of American resources in the Arctic therefore remains uncertain, but its military presence in the Arctic will not change and it will continue to oppose the military actions of Russia and China in the region.

Since Donald Trump left office, Arctic regional affairs have continued to attract the attention of Joe Biden and his administration. United States Secretary of State Antony Blinken recently emphasized the need for the Arctic region to remain an area of "peaceful cooperation and collaboration." In the meantime, a military agreement was reached between the United States military and Norway over building facilities on their military bases, and NATO military exercises in the region have increased. However, Burzynska-Hernandez[28] argued that the United States still lacks leadership in Arctic affairs, which expands the opportunities for Russia and China. Although the United States is politically active in relation to the Arctic, its activities are mostly in response to its competitors, mainly China and Russia. Meanwhile, the adoption of a wait-and-see attitude means that there is a disconnect between politics and the military in the leadership of the United States in the region: "While U.S. military

[25] Hejny, J. (2018). The Trump Administration and environmental policy: Reagan redux? *Journal of Environmental Studies and Sciences, 8*(2), 197–211.

[26] Van de Ven, D.J., Westphal, M., Gonzalez-Eguino, M., Gambhir, A., Peters, G., Sognnaes, I., McJeon, H., Hultman, N., Kennedy, K., & Cyrs, T. (2021). The impact of US re-engagement in climate on the Paris targets. *Earth's Future, 9*(9).

[27] The Biden plan for a clean energy revolution and environmental justice. Biden & Harris.

[28] Burzynska-Hernandez, O. (2021). Action in the Arctic: Lack of US leadership expands opportunities for Russia and China. *Climate Security Risk Briefers,* 42.

and intelligence actors have taken notice of the rapid geopolitical developments in the Arctic, U.S. government officials have been slightly slower on the uptake."

5.2   THE UNITED STATES AND RUSSIA
IN THE ARCTIC: RIVALRY AND DANGERS

From the mid-eighteenth century to the end of the Cold War, relations between the United States and the Soviet Union/Russia in the Arctic have alternated between cooperation and confrontation.[29] At times when they have had common external enemies, the Arctic has been a useful site for cooperation and serving common interests. For example, when they fought as allies in WWII, the Arctic was used as a shortcut for logistics and delivering aid. However, at times when they have been competing in the region or globally, the Arctic has been a bridgehead for military intervention or strategic deterrence. During the Cold War, the Arctic had strategic importance given its geographic proximity to both the United States and the Soviet Union.[30] It was the best location with the shortest range for a strike by either side. In this period, the two rivals were therefore more interested in the use of military force in the Arctic than in the Arctic itself. Since the collapse of the Soviet regime, Russia has been too weak to challenge the United States. Despite nuclear deterrence still being the core security strategy of both the United States and Russia, efforts were made to reduce the number of nuclear weapons in the Arctic region in the 1990s. Combined with other multilateral agreements, such as the Arctic Environmental Protection Strategy and the Arctic Council, this led to a period of cooperation that is referred to as the era of Arctic Exceptionalism.

The annexation of Crimea by Russia in 2014 led to a deterioration in Russia's relations with the United States and European countries, which had a negative impact on cooperation in the Arctic.[31] The United States canceled its joint search-and-rescue exercises with Russia and other joint

[29] Beixi, D. (2016). *Arctic geopolitics*. Arctic.

[30] Huebert, R. (2019). *Breaking the ice curtain? Russia, Canada, and Arctic security in a changing circumpolar world* (pp. 75–93).

[31] Nikulin, M. (2021, February). The Arctic as a potential space for Great Power Competition. In *IOP Conference Series: Earth and Environmental Science, 678*(1), 012,034. IOP Publishing.

projects on mineral deposits and research were also affected. Economic sanctions were also imposed on Russia by the United States and the European Union, targeting Russia's economic vulnerability stemming from its future reliance on offshore oil and gas extraction in the Arctic.[32] The energy sector plays an important role in Russia's economy. Although oil and gas are currently extracted from land fields, Russia's energy strategy is heading toward the tapping of offshore oil and gas reserves in the Arctic. However, Russia's dependence on Western investment and technology for offshore extraction projects made it vulnerable to the sanctions that were imposed.

The Crimean crisis also led to the suspension of most forms of military cooperation, adding more risks to the increasingly visible security dilemma as Russia and the NATO Arctic states continue to build up their military capabilities in the Arctic. Since 2014, Russian military capability in the Arctic has developed significantly to protect Russian interests in the Arctic and the NSR.[33] From 2014 to 2019, Russia acquired 475 facilities along the coast of the Arctic Ocean and upgraded its military weapons and systems. The United States followed suit by enhancing its military presence in the region and recognizing the Arctic as its first line of defense. However, Russia's greater number of icebreakers give it an advantage over the United States. Each party tends to respond to military exercises by the other by conducting exercises of similar strength.[34] For instance, in response to the temporary deployment by the United States of Air Force B-1B bombers in Norway, Russia mobilized air force fighters and bomber aircraft and conducted a missile testing exercise to the north of Norway. With Russia no longer participating in the Arctic Security Forces Roundtable and the suspension of the Northern Chiefs of Defense Conference, the Arctic states have no military forum to communicate on security issues. Rising tensions and the lack of military-to-military dialogue may result in unwanted conflicts triggered by miscalculations or misunderstandings.

---

[32] Olesen, M. R. (2017). *Arctic rivalries: Friendly competition or dangerous conflict?* (DIIS Working Paper No. 2017: 06).

[33] Sliwa, Z., & Aliyev, N. (2020). Strategic competition or possibilities for cooperation between the United States and Russia in the Arctic. *The Journal of Slavic Military Studies, 33*(2), 214–236.

[34] Wither, J. K. (2021). *An Arctic security dilemma: Assessing and mitigating the risk of unintended armed conflict in the High North.* European Security (pp. 1–18).

Although it is a global military power, the United States does not regard the Arctic region as a key strategic issue. During Obama and Trump's presidencies, the United States tried only to maintain its strength in the Arctic without pursuing the complete suppression of Russia.[35] The *Business Insider* recently published an article pointing out that the true position of the United States in the Arctic is somewhat embarrassing, with only one heavy icebreaker, *USCGC Polar Star* (WAGB-10), which belongs to the Coast Guard and was built 45 years ago. It is reported that the ship's equipment is seriously outdated and many parts can no longer be produced.[36] The United States, aware of its shortage of naval armaments in the high latitudes of the Arctic, has decided to bet on its air force and air defense capabilities and its anti-missile systems. For these purposes, the North American Aerospace Defense Command (NORAD) will play a key role. NORAD is a military organization established in 1958 by the United States and Canada to jointly defend North American airspace. In a video conference with Canadian Prime Minister Trudeau in February 2021, President Biden proposed to increase the defense budget to modernize the North American air defense system and strengthen the Arctic defense satellite and radar network. According to expert assessments, upgrading the North Warning System (NWS) may require an investment of US$15 billion, of which the United States will invest 9 billion.[37] Observers have pointed out that Russia and the United States, which are now experiencing the coldest period of their relations since the end of the Cold War, continue to play a fierce game in the freezing Arctic Ocean.[38,39]

[35] Bouffard, T. J., & Rodman, L. L. (2021). US Arctic security strategies: Balancing strategic and operational dimensions. *The Polar Journal*, 1–28.

[36] Woody, C. (2021, March 11). *The heavy-duty ship the US needs to protect its thawing border with Russia "is just falling apart,"* captain says. Business Insider.

[37] Gilmour, J. G. (2021). *NORAD: Renewal of the North warning system by Canada—or Not?* 7.

[38] Trenin, D. (2020). It will get worse before it gets worse. *Horizons: Journal of International Relations and Sustainable Development, 17*, 86–93.

[39] Urpelainen, J., & Van de Graaf, T. (2018). United States non-cooperation and the Paris agreement. *Climate Policy, 18*(7), 839–851.

5.3 The Contest in the Arctic: An Extension of the Rising Sino–American Competition

Before the election of Trump, China and the United States had moderate relations in the Arctic.[40] When the United States took over as chair of the Arctic Council in 2015, the first China–U.S. Arctic Policy Workshop was held in Shanghai and plans were made for annual meetings, with the next scheduled for Washington in 2016. At that time, the United States had not yet developed a negative attitude toward the Chinese presence in the Arctic. At the same time, China considered the United States as neither a partner nor an opponent to its Arctic strategy, out of concerns that the latter's argument for freedom of navigation in the Arctic may spill over into China's home seas. After the Trump administration was elected, it not only provoked a bilateral trade conflict with China for long persistent current account imbalances but also sought to set up a trade siege on China on multiple fronts.[41,42] This also spreads the competition to the Arctic region. The Trump administration's emphasis on the security of American Arctic interests has continued into the Biden administration. After Biden was elected, the new administration sought to strengthen the position and influence of the United States in the Arctic Council and establish the place of the existing Arctic order within the international order led by the United States. As the United States has started to rely more on its allies for Arctic governance, and as it draws in new allies, it may use these alliances to launch an Arctic confrontation campaign against China and Russia aimed at strengthening its own position and suppressing those of its rivals.[43]

The attitude of the United States toward Chinese activities in the Arctic took a sharp turn in May 2019, when then Secretary of State Mike Pompeo proclaimed in his speech in Finland that the Arctic was an arena

---

[40] Peng, J., & Wegge, N. (2015). China's bilateral diplomacy in the Arctic. *Polar Geography, 38*(3), 233–249.

[41] Sheng, L. (2014b). Economic structure, cost outsourcing and global imbalances. *Journal of Australian Political Economy, 74*, 81–95.

[42] Sheng, L. (2015). Theorizing income inequality in the face of financial globalization. *The Social Science Journal, 52*(3), 415–424.

[43] Bouffard, T., Greaves, W., Lackenbauer, P. W., & Teeple, N. (2020). North American Arctic security expectations in a new US administration. *NAADSN Strategic Perspectives*, 23.

for global power and competition. He denied China's claims of being a "near-Arctic state," insisting that there are only Arctic states and non-Arctic states. His rhetoric was followed by action, as the United States started to pressure Arctic states to join its anti-China movement. For example, the American and Danish governments intervened to prevent Chinese firms from refurbishing three airports in Greenland. The United States was worried about its nearby Thule Air Force Base being threatened by Chinese investment in projects that were suspected to have both economic and military purposes. The Chinese threat campaign launched by the United States had finally reached the Arctic, and other Arctic littoral states were pressured to choose sides.[44],[45] In the same month as Pompeo's speech, the U.S. Department of Defense submitted its Annual Report to Congress on the Military Power of the People's Republic of China. This edition of the report added a topic of China's endeavors in the Arctic, mentioning the Arctic region 21 times and focusing on the potential for future Chinese military expansion in the region. The Department of Defense report describes the Chinese threat theory by pointing to China's current military capabilities, including nuclear submarines, and expressing its wariness of the threat of Chinese submarines in the Arctic.[46]

China put forward its plan for a Polar Silk Road in its Arctic policy white paper released in 2018. However, the United States has always publicly called on other countries to be vigilant against China's BRI, warning that this ambitious vision spans Asia and Europe and involves infrastructure projects that will plunge many countries into corruption and debt and leave them politically subjected to China. Pompeo (2019) argued that the Arctic is full of opportunities and wealth and that the Arctic waterway has the potential to become the new Suez Canal in the

---

[44] Conley, H. A., Tsafos, N., & Williams, I. (2020). *America's Arctic moment: Great power competition in the Arctic to 2050*. Center for Strategic and International Studies.

[45] Heurlin, B. (2019). China-US confrontations in the Arctic region: Strategies and policies. *Asian Studies International Journal, 1*(1), 8–15.

[46] United States Navy. (2019). *Strategic outlook for Arctic*, 14.

twenty-first century.[47] The United States seems to be copying its practices in the South China Sea to confront China in Arctic affairs.[48]

China also seems ready to use its economic leverage to influence the position of other Arctic states in Sino–American competition, such as over the issue of Huawei 5G infrastructure.[49] As part of this competition over 5G rollout, the United States has been warning European states over accepting Huawei 5G networks. In an audio recording of a conversation between China's Ambassador to Denmark and the Prime Minister of Denmark, the Chinese Ambassador linked the signing of the China–Denmark free trade agreement to Huawei receiving 5G network contracts in Europe.[50] America's treatment of Russia and China as competitors had led the two countries to cooperate in many areas, including the Arctic.[51] Without territorial claims, China's presence in the Arctic is greatly reliant on its cooperation with Arctic states. China's need for energy resources and Russia's need for investment generate a strategic convergence. Although Sino–Russian relations are not without disagreements, their common perception of the need to counteract American pressure strengthens the ties between the two powers.

5.3.1   The Perception of the United States on China's Polar Silk Road and BRI

Close attention has been paid to China's participation in Arctic affairs and continue to influence the country's decision-making on Arctic strategy toward China. Some scholars have emphasized that China has three core interests in the Arctic region. First, it has security interests, which encompass both traditional and non-traditional security issues. Second,

---

[47] Pompeo, M. R. (2019). *Looking north: Sharpening America's Arctic focus*. US Department of State, 6.

[48] Antsygina, E., Heininen, L. L. M., & Komendantova, N. (2020). A comparative study on the cooperation in the Arctic Ocean and the South China Sea. In *The Arctic. Current issues and challenges* (pp. 83–107).

[49] Patey, L. (2020). *Managing US-China rivalry in the Arctic: Small states can be players in great power competition*. Danish Institute for International Studies.

[50] Kruse, S., & Winther, L. (2019, December 10). *Banned recording reveals China ambassador threatened Faroese leader at secret meeting*. Berlingske.

[51] Saxena, A. (2021, March 19). *The return of Great power competition to the Arctic*. The Arctic Institute.

it has resource interests. Access to the energy and mineral resources in the Arctic region, including exploitable oil, natural gas, fishery, and tourism resources, can provide an important guarantee for China's future economic growth. Third, it has strategic interests, in that the acceleration of China's participation in Arctic affairs is conducive to expanding its global influence and enhancing its capacity for global action. The Polar Silk Road initiative further promotes China's participation in Arctic affairs and enhances its influence over those affairs.[52,53]

The economic drivers of the BRI are clear, and if implemented successfully it can resolve many of China's economic concerns. However, many American observers perceive the BRI as a Chinese version of the Marshall Plan.[54] The BRI is paired with the Marshall Plan by several common features: both intend to resolve their problem of overcapacity by access to overseas markets; both levitate the international status of their currency; both use their abundant foreign reserves to invest overseas; and both involve overseas infrastructure construction plans. The United States is concerned that the BRI is part of China's economic statecraft as it aims to gain great power status and compete with the United States at the international level.[55] BRI loans to developing countries can possibly extend China's political influence and place vulnerable regions like Africa at risk of "neo-colonialism." The BRI has become a major source of infrastructure investment for Indo-Pacific countries. As of 2018, China had provided the region with investments totaling US$90 billion, projects worth US$400 billion, and significant technological support.[56] The economic statecraft can be reflected in voting patterns at the United Nations. For example, in votes on China's Xinjiang and Hong Kong policies, China received supporting votes from 50 and 54

---

[52] Brady, A.-M. (2019). Facing up to China's military interests in the Arctic. *China Brief*, *19*(21).

[53] Tillman, H., Yang, J., & Nielsson, E. T. (2018). The polar silk road: China's new frontier of international cooperation. *China Quarterly of International Strategic Studies*, *4*(3), 345–362.

[54] Li, J. Y. (2020). The change of Western cognition of "the Belt and Road" and China's guiding strategy. Modern Economics & Management Forum.

[55] Hossain, K. (2019). China's BRI expansion and great power ambition: The Silk Road on the ice connecting the Arctic. *Cambridge Journal of Eurasian Studies*, *3*.

[56] Han, Z., & Paul, T. V. (2020). China's rise and balance of power politics. *The Chinese Journal of International Politics*, *13*(1), 1–26.

countries, respectively, with 22 and 39 countries registering criticism of China's policies. Most of the supporting votes were from African, Middle Eastern, and Asian countries with strong economic ties to China.[57]

The United States also speculates on China's strategic rationale for the BRI. China's westward development strategy was first raised in the 1990s out of concerns over a blockade imposed by the United States using its maritime allies along China's east coast.[58] The idea again gained popularity when China became concerned over Obama's Pivot to Asia and over the TPP as obstruction to China's economic growth.[59] By investing in building ports, China will obtain alternatives to the Straits of Malacca, which are controlled by the Navy, thus evading the so-called Malacca Strait Dilemma. For example, the project in Kyaukphyuon can transport 440,000 barrels a day to Yunnan from the Middle East without passing through the Straits of Malacca.[60] The United States is wary that the BRI has military implications, with the Chinese-owned ports in Pakistan, Sri Lanka, Bangladesh, and Myanmar reflecting the pattern of the Chinese "String of Pearls" military strategy.[61] The blurred distinction between civilian and military interests suggests possible military uses for the Chinese commercial ports built under the BRI. Furthermore, the Arctic region is rich in resources and could in future become the shortest shipping route from Asia to Europe. Therefore, China's BRI has a significant impact on the economic ties between China and Europe. The United States sees the BRI as a means for China to expand militarily, and cites China's activities in the Indian Ocean, where it seeks overseas bases and looks to expand its influence, as an example of Chinese expansionism.

The growing presence of China in the Arctic is increasingly convincing the United States that China is working its way to becoming a polar great

---

[57] Parepa, L. A. (2020). The Belt and Road Initiative as continuity in Chinese foreign policy. *Journal of Contemporary East Asia Studies, 9*(2), 175–201.

[58] Rolland, N. (2017). China's "Belt and road initiative": Underwhelming or game-changer?. *The Washington Quarterly, 40*(1), 127–142.

[59] Li, M. (2020b). The Belt and Road Initiative: geo-economics and Indo-Pacific security competition. *International Affairs, 96*(1), 169–187.

[60] Len, C. (2015). China's 21st Century Maritime Silk Road initiative, energy security and SLOC access. *Maritime Affairs: Journal of the National Maritime Foundation of India, 11*(1), 1–18.

[61] Myers, L. (2021, July 20). Internal politics, instability, and China's frustrated efforts to escape the "Malacca Dilemma". Wilson Center.

power.[62] The United States should be concerned that China's current economic presence in the Arctic may transform into a military presence. For navigating the Arctic, China has developed its own icebreaker and is planning to build nuclear-fueled icebreakers, which only Russia currently has in its inventory. China's scientific research stations in the Arctic are assisting it with resource extraction in the region and preparing it for operating in the Arctic's extreme climate: the Xuelong icebreaker, for example, has helped China to acquire useful operational experience by making 10 scientific expeditions. As China's economic interests in the Arctic grow, China may deploy nuclear submarines to Arctic waters to secure shipping routes along the Polar Silk Road, which is a scenario that the United States would not favor.[63] The United States should also beware of the potential for further development of the Polar Silk Road to promote China's great power status outside of the Arctic region.[64] Since China has announced the building of a Polar Silk Road, it has actively built bilateral relationships with Arctic states (other than the United States) by providing large capital investments. For example, the Chinese Arctic and Antarctic Administration launched the Arctic Environment Satellite and Numerical Weather Forecasting Project, under which China has built satellite stations in a number of Arctic states and is planning to establish more. These stations can be potentially used by China to control space systems and access mission data, which would help to project China's space power.

5.4   SUMMARY

Despite distance and desolation, the region around the North Pole has been an arena for the game between major powers for the past century, from WWII and through the Cold War to the present. As the key player of the game in the North Pole, the Arctic policy of the Unites States has undergone a series of adjustments in this period in accordance with

[62] Rush, D., Alexis, D., H., & Gaoqi, Z. (2021). *Northern expedition: China's Arctic activities and ambitions*. Foreign Policy at Brookings.

[63] Musto, R. A. (2019). Antarctic arms control as past precedent. *Polar Record, 55*(5), 330–333.

[64] Robinson, J. (2020). Arctic space challenge for NATO emerging from China's economic and financial Assertiveness. *JAPCC Journal*, issue 30.

the international and regional situation and changes in the goals of environmental protection, resource exploitation, increasing employment, and national security. From the administration of Barak Obama to that of Donald Trump, the focus of the United States on the Arctic shifted from environmental protection to security issues. At the same time, China's increasing presence in the Arctic greatly alarmed the United States. Therefore, the regional situation became fierce once again with a range of military movements by the United States. At present, the Biden administration has retained some of Donald Trump's Arctic policy and is paying attention to both security and environmental issues in the region. In these circumstances, the situation of the North Pole still hangs in the balance.

References

Allison, G. (2018). *China and Russia: A strategic alliance in the making*. The National Interest report. https://nationalinterest.org/feature/china-and-rus sia-strategic-alliance-making-38727

Antsygina, E., Heininen, L. L. M., & Komendantova, N. (2020). A comparative study on the cooperation in the Arctic Ocean and the South China Sea. In *The Arctic: Current issues and challenges*(pp. 83–107). Nova.

Arlota, C. (2018). *Cost & benefit analysis of the United States' withdrawal from the Paris Agreement* (GNLU Working Paper). https://papers.ssrn.com/sol3/papers.cfm?abstract_id=3840390

Beixi, D. (2016). Arctic geopolitics. The impact of US-Russian relations on Chinese Russian cooperation in the Arctic. *Russia in Global Affairs, 14*(2), 206–220.

Bouffard, T. J., & Rodman, L. L. (2021). US Arctic security strategies: Balancing strategic and operational dimensions. *The Polar Journal, 11*, 1–28.

Bouffard, T., Greaves, W., Lackenbauer, P. W., & Teeple, N. (2020). *North American Arctic Security Expectations in a New US administration*. NAADSN Strategic Perspectives. https://www.naadsn.ca/wp-content/uploads/2020/11/North-American-Arctic-Security-Expectations-in-a-New-U.S.-Administr ation-Final.pdf

Brady, A.-M. (2019). Facing up to China's military interests in the Arctic. JamesTown Foundation China Brief. https://www.realcleardefense.com/art icles/2019/12/11/facing_up_to_chinas_military_interests_in_the_arctic_114 913.html

Burzynska-Hernandez, O. (2021). Action in the Arctic: Lack of US leadership expands opportunities for Russia and China. *Climate Security Risk Briefers*, 42–46. https://climateandsecurity.org/wp-content/uploads/2021/10/Cli

mate-Security-Risk-Briefers_Climate-and-Security-Fellows-Program_October-2021-1.pdf

Conley, H. A., & Melino, M. (2019). *The implications of US policy stagnation toward the Arctic region.* Center for Strategic & International Studies report. https://www.csis.org/analysis/implications-us-policy-stagnation-toward-arctic-region

Conley, H. A., Tsafos, N., & Williams, I. (2020). *America's Arctic moment: Great power competition in the Arctic to 2050.* Center for Strategic and International Studies report. https://www.csis.org/analysis/americas-arctic-moment-great-power-competition-arctic-2050

Doshi, R., Dale-Huang, A., & Zhang, F. Q. (2021). *Northern expedition: China's Arctic activities and ambitions.* Brookings China Strategy Initiative report. https://www.brookings.edu/research/northern-expedition-chinas-arctic-activities-and-ambitions/

Gilmour, J. G. (2021). *NORAD: Renewal of the North Warning System by Canada or not?* Naval Association of Canada report. https://www.navalassoc.ca/wp-content/uploads/2021/08/Gilmour-NORAD.pdf

Han, Z., & Paul, T. V. (2020). China's rise and balance of power politics. *The Chinese Journal of International Politics, 13*(1), 1–26.

Harball, E. (2018). Trump administration approves first oil production in federal Arctic waters. https://www.alaskapublic.org/2018/10/24/trump-administration-approves-first-oil-production-in-federal-arctic-waters/.

Haycox, S. (2020). Arctic policy of the United States: An historical survey. In *The Palgrave Handbook of Arctic Policy and Politics*, 233–250. Palgrave Macmillan.

Hejny, J. (2018). The Trump Administration and environmental policy: Reagan redux? *Journal of Environmental Studies and Sciences, 8*(2), 197–211.

Heurlin, B. (2019). China-US confrontations in the Arctic region: Strategies and policies. *Asian Studies International Journal, 1*(1), 8–15.

Hossain, K. (2019). China's BRI expansion and great power ambition: The Silk Road on the ice connecting the Arctic. *Cambridge Journal of Eurasian Studies.* https://research.ulapland.fi/en/publications/chinas-bri-expansion-and-great-power-ambition-the-silk-road-on-th

Huebert, R. (2009). *United States arctic policy: The reluctant arctic power.* The School of Public Policy–University of Calgary Publications Series, 2(2). https://papers.ssrn.com/sol3/papers.cfm?abstract_id=3053702

Huebert, R. (2019). *Breaking the ice curtain? Russia, Canada, and Arctic security in a changing circumpolar world*, 75–93. https://d3n8a8pro7vhmx.cloudfront.net/cdfai/pages/4193/attachments/original/1558816637/Breaking_the_Ice_Curtain.pdf?1558816637

Käpylä, J., & Mikkola, H. (2019). Contemporary Arctic meets world politics: Rethinking Arctic exceptionalism in the age of uncertainty. In *The Global Arctic Handbook* (pp. 153–169). Springer.

Kruse, S., & Winther, L. (2019). *Banned recording reveals China ambassador threatened Faroese leader at secret meeting.* Berlingske report. https://www.berlingske.dk/internationalt/banned-recording-reveals-china-ambassador-threatened-faroese-leader.

Lavorio, A. (2021). Geography, climate change, national security: The case of the evolving US Arctic strategy. *The International Spectator, 56*(1), 111–125.

Len, C. (2015). China's 21st century maritime silk road initiative, energy security and SLOC access. *Maritime Affairs: Journal of the National Maritime Foundation of India, 11*(1), 1–18.

Li, J., Y. (2020). *The change of western cognition of "the belt and road" and China's guiding strategy.* Modern Economics & Management Forum report. https://doi.org/10.32629/memf.v1i3.240

Li, M. (2020). The Belt and Road Initiative: Geo-economics and Indo-Pacific security competition. *International Affairs, 96*(1), 169–187.

Mariia, K. (2019). China's Arctic policy: Present and future. *The Polar Journal, 9*(1), 94–112.

McCann, K. B., McMullin, J. A., & Turvold, W. D. (2020). *Before it's too late: US Maritime Grand Strategy in the Inidan Ocean and the Arctic Ocean.* Daniel K. Inouye Asia-Pacific Center for Security Studies report. https://apcss.org/nexus_articles/before-its-too-late-u-s-maritime-grand-strategy-in-the-indian-ocean-and-the-arctic-ocean/

Musto, R. A. (2019). Antarctic arms control as past precedent. *Polar Record, 55*(5), 330–333.

Myers, L. (2021). *Internal politics, instability, and China's frustrated efforts to escape the "Malacca Dilemma".* Wilson Center report. https://www.wilsoncenter.org/blog-post/internal-politics-instability-and-chinas-frustrated-efforts-escape-malacca-dilemma.

Nikulin, M. (2021). The Arctic as a potential space for great power competition. In *IOP Conference Series: Earth and Environmental Science.* IOP Publishing. https://doi.org/10.1088/1755-1315/678/1/012034

Nilsson, A. E. (2018). The United States and the making of an Arctic nation. *Polar Record, 54*(2), 95–107.

Olesen, M. R. (2017). *Arctic rivalries: Friendly competition or dangerous conflict?* (DIIS Working Paper, 2017/6). https://pure.diis.dk/ws/files/1141975/DIIS_WP_2017_6.pdf

Parepa, L. A. (2020). The belt and road initiative as continuity in Chinese foreign policy. *Journal of Contemporary East Asia Studies, 9*(2), 175–201.

Patey, L. (2020). *Managing US-China rivalry in the Arctic: Small states can be players in great power competition.* Danish Institute for International

Studies report. https://www.diis.dk/en/research/managing-us-china-rivalry-in-the-arctic.

Peng, J., & Wegge, N. (2015). China's bilateral diplomacy in the Arctic. *Polar Geography, 38*(3), 233–249.

Pompeo, M. R. (2019). *Looking North: Sharpening America's Arctic focus.* US Department of State speech. https://2017-2021.state.gov/looking-north-sharpening-americas-arctic-focus/index.html

Robinson, J. (2020). *Arctic space challenge for NATO emerging from China's economic and financial assertiveness.* Joint Air Power Competence Centre report. https://www.japcc.org/arctic-space-challenge-for-nato/

Rolland, N. (2017). China's "Belt and Road Initiative": Underwhelming or game-changer? *The Washington Quarterly, 40*(1), 127–142.

Saxena, A. (2021). *The return of great power competition to the Arctic.* The Arctic Institute report. https://www.thearcticinstitute.org/return-great-power-competition-arctic/

Sheng, L. (2014a). Capital controls and international development: A theoretical reconsideration. *Global Policy, 5*(1), 114–120.

Sheng, L. (2014b). Economic structure, cost outsourcing and global imbalances. *Journal of Australian Political Economy, 74*, 81–95.

Sheng, L. (2015). Theorizing income inequality in the face of financial globalization. *The Social Science Journal, 52*(3), 415–424.

Sheng, L. (2021). *How Covid-19 reshapes new world order: Political economy perspective.* Springer.

Sheng, L., & do Nascimento D. F. (2021a). *The belt and road Initiative in South-South cooperation: The impact on world trade and geopolitics.* Palgrave MacMillan.

Sheng, L., & do Nascimento, D. F. (2021b). *Love and trade war: China and the US in historical context.* Springer.

Sheng, L., Li, T., & Wang, J. (2017). Tourism and externalities in an urban context: Theoretical model and empirical evidence. *Cities, 70*, 40–45.

Sheng, L., & Zhao, W. (2016). Strategic destination management in the face of foreign competition: The case of Macao SAR. *Journal of Travel & Tourism Marketing, 33*(2), 263–278.

Sliwa, Z., & Aliyev, N. (2020). Strategic competition or possibilities for cooperation between the United States and Russia in the Arctic. *The Journal of Slavic Military Studies, 33*(2), 214–236.

Tillman, H., Yang, J., & Nielsson, E. T. (2018). The polar silk road: China's new frontier of international cooperation. *China Quarterly of International Strategic Studies, 4*(3), 345–362.

Trenin, D. (2020). It will get worse before it gets worse. *Horizons: Journal of International Relations and Sustainable Development, 17*, 86–93.

Urpelainen, J., & Van de Graaf, T. (2018). United States non-cooperation and the Paris agreement. *Climate Policy, 18*(7), 839–851.

United States Navy. (2019). *Strategic outlook for Arctic.* https://media.defense.gov/2020/May/18/2002302034/-1/-1/1/NAVY_STRATEGIC_OUT LOOK_ARCTIC_JAN2019.PDF.

Vakhshouri, S. (2017). *America First Energy Plan: Renewing the confidence of American energy producers.* Atlantic Council report. https://www.atl anticcouncil.org/in-depth-research-reports/issue-brief/america-first-energy-plan/#:~:text=In%20an%20issue%20brief%20on%20US%20energy%20product ion,to%20part%20to%20growing%20support%20for%20climate%20policy.

Van de Ven, D.J., Westphal, M., Gonzalez-Eguino, M., Gambhir, A., Peters, G., Sognnaes, I., McJeon, H., Hultman, N., Kennedy, K., & Cyrs, T. (2021). The impact of US re-engagement in climate on the Paris targets. *Earth's Future, 9*(9), e2021EF002077.

Wither, J. K. (2021). An Arctic security dilemma: Assessing and mitigating the risk of unintended armed conflict in the High North. *European Security, 30*(4), 649–666.

Woody, C. (2021). *The heavy-duty ship the US needs to protect its thawing border with Russia "is just falling apart," captain says.* Business Insider Nederland report. https://www.businessinsider.com/coast-guard-icebreaker-polar-star-running-out-of-spare-parts-2021-3.

Yin, Y. C., & Sheng, L. (2021). Theorizing about global imbalances: An inequality perspective. *Argumenta Oeconomica, 46*, 169–181.

CHAPTER 6

Who Will Win in the Climate Crisis? A Reinterpretation of the Interaction Between Climate Change and Security Issues in the Arctic

Abstract This chapter analyzes how the major Arctic powers are responding to the conventional and unconventional challenges posed by climate change in the Arctic. Climate change in the Arctic is a double-edged sword for Russia, which faces the challenges of permafrost degradation and reduced food production but also opportunities from the opening of the Arctic shipping route. Meanwhile, growing Sino–Russian cooperation in the Arctic has given Russia an additional bargaining chip in the Arctic game with the United States.

Keywords Arctic competition · Climate change · Global warming · Sino–Russian relations · International cooperation

Despite being on the geographic periphery of the globe, the Arctic is capable of exerting great influence over the rest of the world. As the melting of the Arctic ice caps accelerates due to global warming, events in this remote and freezing region are spilling over to affect the whole world in a range of areas, including the global climate and the landscape of international navigation and energy supplies. These changes in the Arctic will confront countries around the world with a series of unconventional

© The Author(s), under exclusive license to Springer Nature Singapore Pte Ltd. 2022
E. L. Sheng, *Arctic Opportunities and Challenges*,
https://doi.org/10.1007/978-981-19-1246-7_6

challenges. On the one hand, the world will enjoy the convenience of shortened shipping routes and more plentiful supply of resources. On the other hand, the whole international community must manage the risks of natural disasters and loss of habitat as a result of global warming. In this chapter, we mainly discuss how the three major Arctic powers are responding to the conventional and unconventional challenges posed by changes in the Arctic.

6.1   RUSSIA: THE MAIN BENEFICIARY OF GLOBAL WARMING?

Russia, as a northern Eurasian country and the largest Arctic country, has been a pioneer in exploring the North Pole, for two main reasons. First, Russia depends on the rich natural resources in its Arctic territory, such as oil and gas, to boost its domestic economy and employment. Second, Russia looks to strengthen its internal transportation from its developed western regions to its remote Far East and east coast by developing maritime routes in the Arctic Ocean. However, it is difficult for Russia to independently realize its ambitions in its Arctic territory, for many reasons. First, the crises in Ukraine and Crimea have made it difficult for Russia to fully concentrate on its northern frontier, considering that the Arctic is still the subject of territorial disputes. Second, having suffered from weak economic performance for many years, the lack of financial and technical investments hinders the further development of the region by Russia. Third, labor deficits, especially in the Far East but also across the whole country, place constraints on carrying out large-scale projects. For these reasons, the Polar Silk Road could bring great benefits to Russia in developing its Arctic territory. Seeking international partners such as China for these endeavors is undoubtedly a top priority as it can not only promote Russia's domestic economic development but also improve its extraction technology and local labor force, especially in the Far East.[1] This chapter discusses the cooperation between China and Russia in the Arctic from the perspective of climate change.

With 1.71 gigatons of CO_2 emissions, Russia is the fourth-largest emitter of greenhouse gases in the world, behind China, the United

---

[1] Sheng, L., & do Nascimento, D. F. (2021a). *The belt and road initiative in south-south cooperation: The impact on world trade and geopolitics*. Palgrave MacMillan.

States, and India.[2] Russia is rich in natural resources and its economy is highly dependent on energy exports. Although Russia's gas and oil production decreased substantially in 2020, it remains the world's largest net exporter of oil and gas combined.[3] The massive extraction of oil undoubtedly has catastrophic consequences for the climate and the environment. As the Natural Gas Exporting Countries Forum (GECF) pointed out, the carbon dioxide produced by natural gas combustion is only half that of coal, and the use of gas has made an indispensable contribution to protecting the environment and especially to mitigating and adapting to climate change.[4] However, the export of natural gas may lead to higher emissions in exporting countries, thus having the opposite results. For exporting countries, the export of liquefied natural gas must go through an energy-intensive process of liquefaction, transportation, and regasification. The emissions intensity of natural gas exports is therefore higher than that of the domestic production and consumption of natural gas, with most of these higher emissions coming from liquefaction stations in exporting countries. High quantities of methane can also leak or be vented into the atmosphere during the drilling, transportation, and processing of the gas.[5] Russia therefore faces great challenges related to climate change.

The temperature in Russia is rising faster than the global average, and this will have a range of serious consequences, such as the melting of frozen soil and damage to infrastructure, and lead to subsequent waves of migration. In the 1990s, research showed that increasing annual average air temperatures would result in permafrost degradation in some regions of Russia, which would affect local ecological conditions and the stability of buildings.[6] In the summer of 2020, there were several

[2] Union of Concerned Scientists. (2020). *Each country's share of CO$_2$ emissions*, updated August 12, 2020.

[3] British Petroleum. (2021). *Statistical review of world energy—2021 Russia's energy market in 2020.*

[4] GECF. (2021, February 24). *Global gas outlook 2050.*

[5] Swanson, C., & Levin, A. (2020, December 8). *Liquefied natural gas exports are a climate threat.*

[6] Vyalov, S. S., Gerasimov, A. S., & Fotiev, S. M. (1998, June 23–27). *Influence of global warming on the state and geotechnical properties of permafrost. Permafrost: Proceedings of the Seventh International Conference*, Yellowknife, NW, 1097–1102.

sudden building collapses in Yakutsk, the capital of Siberia.[7] These were caused by climate change, as the increased summer temperatures and greater depth of underground thawing led to the thawing of originally frozen rocks, resulting in the degradation of underground permafrost and building foundations. Meanwhile, the melting of the permafrost causes the microorganisms in the soil to wake up and digest organic matter, thereby converting carbon into the greenhouse gases carbon dioxide and methane and releasing these into the atmosphere, leading to the intensification of climate warming. Climate change is also expected to increase droughts in Russia's rich southern agricultural "bread-basket" regions, including Stavropol and Rostov.[8] Russia has recently become the world's leading wheat exporter, with the International Grains Council (IGC) estimating total Russian grain production for 2020–2021 at 125.6 million tons, up from its previous estimate of 125.1 million.[9] With the increase in temperature, the reproduction and metabolism of pests will speed up, weed will be encouraged to proliferate, and water resource shortages will be aggravated, which is likely to cause a significant reduction in the production of major food crops, such as wheat, corn, barley, and rice, in Russia. Climate change will therefore certainly have a negative effect on Russia's grain production, although it may also bring about new opportunities for Russian agriculture. Kiselev pointed out that global climate change is expected to have an overall positive impact on Russian agriculture through the increase in the growing season and the expansion of agricultural areas.[10]

Given Russia's large greenhouse gas emissions and energy exports, it plays a pivotal role in global climate change governance. Russia joined the Kyoto Protocol in 2004 and the Paris Agreement in 2015. After the adoption of the Paris Agreement by a decree from President Vladimir Putin,

---

[7] Chenguang|Zhuo yan tianxia: Yongjiu dongtu ronghua quanqiu weiji jiaju. [Chenguang|Focus on the world: Melting permafrost global crisis intensifies].

[8] Conley, H. A. (2021). *Climate change will reshape Russia*. Center for Strategic International Studies,

[9] Lyddon, C. (2021, August 8). *Focus on Russia*. World Grain.

[10] Kiselev, S., Romashkin, R., Nelson, G. C., Mason-D'Croz, D., & Palazzo, A. (2013). Russia's food security and climate change: Looking into the future. *Economics, 7*(1).

Russia officially set a goal to reduce greenhouse gas emissions to 70–75% of 1990 emissions by 2030.[11] However, Russia's shaky position in international climate governance and its ineffective measures to stimulate domestic economic upgrades generate skepticism over Russia's commitment. Domestically, the Russian government has faced great pressure from its population to combat climate change. According to a poll, after a heatwave that led to extensive forest, farmland, and bog fires in central European Russia in 2010, the proportion of Russians worried about climate change increased from a pre-2010 figure of 46–55%. By 2013, the proportion in Moscow had risen to 70%.[12] In the 2009 Climate Doctrine and 2011 Climate Action Plan, Russia stated its aims to increase energy efficiency in all sectors of the economy; develop and deploy renewable and alternative energy sources; reduce market imbalances and implement financial and fiscal policies to encourage anthropogenic greenhouse gas emission reduction; and protect and enhance the capacity of carbon sinks, including sustainable forestry, forestation, and reforestation.[13] In 2013, a presidential decree on reducing greenhouse gas emissions was approved that adopted a target of no greater than 75% of the total emissions of 1990 by 2020.[14]

These policies, however, have not been effective in delivering what is needed to combat climate change. According to the assessment of the Climate Action Tracker, Russia's emission reduction strategies are critically insufficient because in its newly adopted Energy Strategy running through to 2035, Russia continues to focus on expanding its domestic production and consumption of fossil fuels, with a strong emphasis on expanding natural gas exports.[15] There are two explanations for Russia's ineffective policies. First, they are driven by economic considerations rather than environmental concerns. The tight control of state-owned

[11] Oshchepkov, M. (2021, July 11). *Russia has set an ambitious goal for reducing emissions by 2030*. Climate Scorecard.

[12] Tynkkynen, V.-P., & Tynkkynen, N. (2018). Climate denial revisited: (Re)contextualising Russian public discourse on climate change during Putin 2.0. *Europe-Asia Studies, 70*(7), 1103–1120.

[13] Sharmina, M., Anderson, K., & Bows-Larkin, A. (2013). Climate change regional review: Russia. *Wiley Interdisciplinary Reviews: Climate Change, 4*(5), 373–396.

[14] Grantham Research Institute on Climate Change and the Environment, *Greenhouse Gas emission reduction* (Presidential Decree 752).

[15] Climate Action Tracker: Russian Federation.

companies and their operations has become detrimental to Putin's ambitions to turn Russia into a global player in the energy industry, with many countries being dependent on Russia's vast oil and gas supplies. Employment and the social infrastructure of the entire region are highly dependent on coal or oil and the infrastructure that energy companies maintain as a legacy of the Soviet era.[16] As a result of Russia's economic dependence on these natural resources, employment opportunities and social infrastructure are largely based on industries involved in the extraction and refinement of oil and gas.

Therefore, Russia has had little incentive to pay attention to climate change, which would eventually necessitate a move away from the resources so important to its economy and society. Second, the Russia government has not yet attached great importance to climate change as an issue, and Putin and other Russian leaders have sporadically flirted with outright global climate change denial. President Putin does not consider the climate agenda other than as a threat to Russia's national security and economy, and the subordination of other branches of power to the chief executive in Russia precludes other actors from developing their own agenda.[17] Similarly, Russia's participation in global international climate governance is driven by political and economic incentives. The European Union, China, and the United States have all made plans and commitments to combat climate change—Europe has adopted a Green Deal, China has committed to be carbon neutral by 2060, and President Biden has replaced climate skeptic Trump in the White House—but Russia has yet to start a transition to carbon neutral sources of energy. Such a transition would however run against the ambitions of Putin to leverage Russia's oil and gas exports in a world that will become less dependent on these resources. Fang Lexian and Wang Yujing pointed that the driving forces for Russia to cooperate with European countries in climate change are promoting domestic economic transformation and developing key technologies.[18]

[16] Champion, M., & Doff, N. (2021, August 15). *Russia's getting left behind in global dash for clean energy*. Bloomberg.

[17] Semenov, A. (2021). *Russian political forces meet climate change*. Center for Strategic International Studies.

[18] Fang Lexian, & Wang Yujing. (2021). *Oumeng yu eluosi qihou hezuo de jinzhan yu juxian. Heping yu Fazhan*. 2021(03).

Because of its geographic position, there is some argument that climate change may bring new opportunities to Russia. For example, global warming could allow for voyages through the sea ice on the NSR, which is controlled by Russia, in winter.[19] Indeed, President Putin remained skeptical about the negative influences of climate change for a long time. In 2003, when asked whether Russia would sign the Kyoto Protocol, Putin replied, "Maybe climate change is not so bad in such a cold country as ours? 2–3 degrees wouldn't hurt—we'll spend less on fur coats, and the grain harvest would go up."[20] At the same time, the melting of the ice on the NSR may bring employment and tourism income to the 5.4 million people living in the harsh Russian Arctic region. In 2017 and 2018, Putin pointed to volcanic eruptions and changes of global character, cosmic changes, or some invisible movements in the galaxy as being responsible for climate change.[21]

As a country situated at a high latitude, global warming will indeed bring some opportunities to Russia, the most important of which will be the easier navigation of the Arctic waterway. However, Russia cannot neglect the effects of the crisis, such as permafrost degradation and a reduction in grain production. Furthermore, the crisis is not only domestic, as changes in the global ecosystem will bring human beings into a completely new situation from which no country is immune.

6.2   The Dilemma of the United States: International Leadership and Domestic Politics

The United Nations Framework Convention on Climate Change (UNFCCC) is at the core of the international system for cooperation on climate change and has played an important role in international agreements on limiting global climate change. The two most influential agreements passed by the UNFCCC are the Kyoto Protocol and the Paris Agreement. As the most powerful state and traditionally viewing itself regard as the "world police," the United States has insisted on a leading

---

[19] Arvin, J. (2021, February 22). *The latest consequence of climate change: The Arctic is now open for business year-round*. Vox.

[20] Skepticism to acceptance: How Putin's views on climate change evolved over the years. *Moscow Times*, July 1, 2021.

[21] Ibid.

position in climate change under the UNFCCC framework. Certainly, the role of the United States in combating global warming is crucial as the world's largest economy and second-largest emitter, with its share of CO_2 emissions of 5.41 gigatons only surpassed by China.[22] However, the attitude of the United States toward multilateral efforts to address climate change has oscillated between engagement and disengagement.[23,24] This section discusses the different policies adopted by American leaders since the Clinton presidency with respect to climate change governance.

The Kyoto Protocol was adopted on December 11, 1997. In short, the Kyoto Protocol operationalizes the UNFCCC by committing industrialized countries and economies in transition to limit and reduce greenhouse gas emissions in accordance with agreed individual targets.[25] The Kyoto Protocol was promoted by Vice President Gore in 1997 and finally signed by President Clinton in 1998.[26] However, Congress failed to ratify the agreement, which means that the United States never officially signed the pact. The Byrd–Hagel resolution, passed five months before the Kyoto meeting in 1997, claimed that unless the protocol mandated a new specific scheduled commitment for developing countries within the same compliance period, it would cause serious losses to the economy and thus the United States should not be a signatory.[27] Research conducted by Hovi, Sprinz, and Bang revealed as a possible explanation for the United States not becoming a party to Kyoto that the Clinton–Gore administration gave up on Senate ratification and essentially pushed for an agreement that would give them a climate-friendly face.[28]

[22] Union of Concerned Scientists. (2020). *Each country's share of CO_2 emissions*, updated August 12, 2020.

[23] Kelemen, D. R., & Vogel, D. (2010). Trading places: The role of the US and EU in international environmental politics. *Comparative Political Studies, 43*(4), 427–456.

[24] Parker, C. F., & Karlsson, C. (2018). The UN climate change negotiations and the role of the United States: Assessing American leadership from Copenhagen to Paris. *Environmental Politics, 27*(3), 519–540.

[25] Kyoto Protocol, United Nations Climate Change.

[26] Signing the Kyoto Protocol, White House Green Building: An exhibit on sustainability. *Presidential Libraries*.

[27] Senate Resolution 98 (Congressional Record, Report No. 105–5412), June 1997.

[28] Hovi, J., Sprinz, D. F., & Bang, G. (2010). Why the United States did not become a party to the Kyoto Protocol: German, Norwegian, and US perspectives. *European Journal of International Relations, 18*(1), 129–150.

President Bush adopted a more antagonistic attitude to the Kyoto Protocol than that of his predecessor and rejected the Protocol after he came into power in 2001. He reiterated the Byrd–Hagel resolution[29] and argued that the Kyoto Protocol exempted China and India from compliance, which would cause serious harm to the economy.[30] Bush's action frustrated America's European allies who were actively promoting the climate change agenda in world politics. A European diplomat called the hardline attitude depressing and described it as marking a significant divergence between European and American views on how to deal with the problem.[31] In February 2002, President Bush committed the United States to a comprehensive strategy to reduce the greenhouse gas emission intensity of its economy by 18% by 2012,[32] which was regarded as a less stringent alternative to the Kyoto Protocol.[33] However, Bush's rejection was seen as a major setback to the success of the Kyoto Protocol and seriously undermined the leadership position of the United States in climate change negotiations.[34] When Obama became president, he sought to restore the country's leadership and adopted a more active approach to multilateral negotiations on climate change. Obama made a renewed commitment to climate change mitigation in 2014, pledging to reduce emissions to 26–28% below 2005 levels within 10 years.[35] In 2015, he insisted on the responsibility of the United States and further pledged to reduce carbon emissions to 17% below 2005 levels by 2020.[36] Through these emission commitments, Obama aimed to emphasize the leadership of the United States in global governance. In his statement on the Paris Agreement, Obama said that the world was safer, more secure,

---

[29] Ibid.

[30] Text of a Letter from the President to Senators Hagel, Helms, Craig, and Roberts, March 2001. White House Archives.

[31] Borger, J. (2001, March 29). Bush kills global warming treaty. *The Guardian*.

[32] *Clean energy and climate change*. Bush White House Archives.

[33] Bush unveils voluntary plan to reduce global warming, February 14, 2002. *CNN*.

[34] Outka, U. (2016). The Obama administration's Clean Air Act legacy and the UNFCCC. *Case Western Reserve Journal of International Law, 48*, 109–125.

[35] *Remarks by President Obama at the first session of COP21*, November 30, 2015. White House Archives.

[36] Ibid.

more prosperous, and much more free because of strong, principled, American leadership.[37]

The election of President Trump jeopardized the continued participation of the United States in the Paris Agreement and the UNFCCC.[38] Trump brought mercantilist concepts to the forefront of America's politics with the slogan of "America First" and increased trade frictions with other countries with large current account surplus.[39] He successively withdrew from the United Nations Educational, Scientific and Cultural Organization (UNESCO), the UNFCCC, and the WTO. Adopting the principle of "all but Obama," he abandoned Obama's health care policy, Asia-Pacific rebalancing strategy, and commitment to multilateral negotiations on climate change. The Trump administration, managed by climate skeptics, eliminated the Obama administration's regulations on greenhouse gas emissions and led the United States to become the first country to officially withdraw from the Paris Agreement.[40] Meanwhile, the Trump administration claimed that the climate plan implemented by the Obama administration was the main reason for the "Coal War"[41] in the United States. As a result, as soon as Trump took office, he committed to overthrowing the climate policy formulated during the Obama era and emphasized that relevant climate policies were damaging the economy of the United States and reducing employment. At the same time, the Trump administration has imposed trade restrictions on China as the U.S. trade deficit with China has widened sharply, albeit ignoring

[37] *Statement by the President on the Paris Climate Agreement*, December 12, 2015. White House Archives.

[38] Parker, C. F., & Karlsson, C. (2018). The UN climate change negotiations and the role of the United States: Assessing American leadership from Copenhagen to Paris. *Environmental Politics, 27*(3), 519–540.

[39] Yin, Y. C., & Sheng, L. (2021). Theorizing about global imbalances: An inequality perspective. *Argumenta Oeconomica, 46,* 169–181.

[40] Sheng, L. (2021). *How Covid-19 reshapes new world order: Political economy perspective.* Springer.

[41] In January 2011, the U.S. Environmental Protection Agency provoked an outcry in West Virginia by rejecting a permit to fill a river valley. West Virginia is America's second-largest coal producing state, and this permission was crucial to the largest mountaintop removal mining operation in the United States.

the underlying structural imbalance between the two countries.[42,43] On January 20, 2017, Trump signed the America First Energy Plan on the day of his inauguration, repealing the Obama administration's Climate Action Act. Trump believed that the United States should immediately return to the traditional coal industry and increase the use of fossil fuels, progressively lift energy restrictions and government intervention, eliminate unnecessary climate change policies, and accelerate the country's economic recovery. To achieve this, Trump also put forward some ambitious plans to revitalize the coal industry. However, the move triggered the joint opposition of 23 U.S. states, local governments, and environmental protection organizations, claiming that these plans would pose a threat to public health and vowing to fight it in court. Moreover, the Trump administration overthrew the Clean Power Plan (CPP) signed by the Obama administration on October 10, 2017, in what was widely considered as a landmark event in the decline of American climate politics.[44] Trump's behavior heavily undermined the international prestige of the United States and destroyed Obama's efforts to re-establish America's leading role in combating climate change.

In the mid-term elections in 2021, Biden defeated Trump to become president. Just hours after being sworn in, Biden moved to reinstate the United States to the Paris Agreement and his administration rolled out a cavalcade of executive orders aimed at tackling the climate crisis.[45] President Biden emphasized the urgency of reinforcing climate action while adopting restrictions on the use of fossil fuels and making large-scale investments in clean energy. President Biden also promised that he would not accept donations from oil, gas, and coal companies or executives.[46] He announced a new target of a 50–52% reduction from 2005 levels in

[42] Sheng, L. (2014b). Economic structure, cost outsourcing and global imbalances. *Journal of Australian Political Economy, 74,* 81–95.

[43] Sheng, L. (2015). Theorizing income inequality in the face of financial globalization. *The Social Science Journal, 52*(3), 415–424.

[44] Christopher, T. (2017, March 28). *Trump signs executive order to roll back Obama-era climate actions, power plant emissions rule.* CNBC.

[45] Milman, O. (2021, January 20). Biden returns US to Paris climate accord hours after becoming president. *The Guardian.*

[46] *The Biden plan for the clean energy revolution and environmental justice.*

economy-wide net greenhouse gas pollution by 2030.[47] Trump's series of retrogressive policies on climate change damaged the credibility and influence of the United States in climate negotiations and hindered the efforts of the international community to reduce greenhouse gas emissions. By returning to the Paris Agreement, the Biden administration was eager to assure the international community that the United States would return to the path of tackling climate change, repair relations with the international community, restore the credibility of the United States in global diplomacy, and regain international prestige. In this way, the United States expects to seize the opportunity to develop new energy industries and boost its economy while returning to its position as a leader in global climate politics.

To conclude, America's attitude to its leadership position on climate change is congruent with its foreign policy tradition of swinging between engagement and disengagement. From President Clinton to President Biden, the United States has had no consistent position in climate change governance. With consideration of domestic and international pressures, the United States continues to seek an equilibrium between domestic economic benefits and international leadership prestige.

6.3   CHINA IN THE ARCTIC: DIPLOMACY OVER THE SILK ROAD AND SINO–RUSSIAN COOPERATION

Climate change is a double-edged sword for Arctic issues. On the one hand, previously frozen parts of the Arctic are opening for navigation, and natural resources under the permafrost are becoming more accessible. On the other hand, global warming is melting Arctic glaciers and rising sea levels, bringing many potential crises to the whole of human society. Thus, Arctic exploitation and the climate crisis require international cooperation. Based on broad common interests, China and Russia have conducted a series of collective actions in the Arctic. The two countries have been hosting regular dialogues, both between their foreign ministries (the China–Russia Dialogue on Arctic Affairs since 2015) and at the expert level (the China–Russia Arctic Forum, co-hosted since 2012

---

[47] FACT SHEET: *President Biden sets 2030 greenhouse gas pollution reduction target aimed at creating good-paying union jobs and securing U.S. leadership on clean energy technologies*, April 22, 2021. White House briefing.

by St. Petersburg State University and the Ocean University of China).[48] On June 5, 2019, China's President Xi Jinping and Russia's President Vladimir Putin decided to upgrade their relations to a comprehensive strategic partnership of coordination for a new era, to symbolize the healthy and stable development of Sino–Russian relations.[49] Sino–Russian Arctic cooperation has since been steadily promoted across the domains of economic, politics, and military affairs. A typical example of bilateral cooperation is the Yamal LNG project, the first large-scale energy cooperation project implemented in Russia after China put forward the BRI.[50] China will use its own strengths in the economy, energy, and other fields as an entry point for closer relations with Russia, promoting the social and economic development of the Russian Arctic region. In doing so, China will also safeguard its Arctic rights and interests.

China and Russia have different considerations in Arctic cooperation. For China, an Ice Sea Route (ISR) forms part of its BRI, with a particular focus on energy security and access to the sea route. China's Arctic cooperation serves its grand security strategies. According to China's 2018 Arctic policy white paper, the ISR is a crucial part of the BRI. China's interest in Arctic issues is driven by the desire for a shorter and more reliable shipping route and access to natural resources, and its insight on climate change.[51] Russia actively supports China's BRI and has reached a multitude of cooperation agreements on Arctic matters, including the development of Arctic resources and the construction of Arctic waterways based on further communication and negotiation, with these efforts having achieved some initial results. Russia needs China's investment to build infrastructure but without Chinese involvement in decision-making

---

[48] Wishnick, E. (2021, June 5). Will Russia put China's Arctic ambitions on ice? *The Diplomat*.

[49] Liang, Y. (2019, June 6). China, Russia agree to upgrade relations for new era. *Xinhua*.

[50] Overseas Project Encyclopedia. *Dujia: Yamal LNG project: bingshang sichouzhilu de zhongyao jiedian* [Yamal LNG project: An important pivot point of the Ice Silk Road, China Belt and Road Network].

[51] Chen, G. (2012). China's emerging Arctic strategy. *The Polar Journal, 2*(2), 358–371.

processes.[52] Russia also needs assistance from China for ecological protection. The protection of the Arctic environment is not only conducive to the stability of the global environment but also favorable to the maintenance of the common interests of global citizens. Especially in a situation such as that of Russia, which has a sluggish domestic economy and is hindered in its local development by European and American sanctions, the active development of international bilateral and multilateral cooperation with other countries is necessary to obtain financial, equipment, and technical support for the social and economic development of the Arctic region.

Zhang Pei and Yang Jian (2016)[53] pointed out four considerations of China in Arctic cooperation: vulnerability to the adverse effects of climate change; expanding overseas energy supplies to safeguard economic security; using the Arctic seaways to explore new areas of economic growth; and depending on Arctic scientific research to build up knowledge reserves and technological innovation for the country's long-term development. Climate change brings China new opportunities in Arctic cooperation, and if China were to adopt a stronger role in mitigating climate change it could legitimate its Arctic role and build a more positive image in the eyes of the other actors in the region.[54] China's experience with Arctic expeditions could also help it to provide intellectual and technical support and strengthen its role in Arctic affairs.

As the Arctic cooperation between China and Russia continues to advance, the relations between the two countries are becoming progressively closer. These close ties can be regarded as a response to America's harsh containment policy toward the two countries, not only in relation to Arctic issues but also other international affairs.[55,56] Although Russia has aggressive territorial ambitions in the Arctic, China does not intend

[52] Pincus, R. (2020). Three-way power dynamics in the Arctic. *Strategic Studies Quarterly, 14*(1), 40–63.

[53] Zhang, P., & Yang, J. (2016). Changes in the Arctic and China's participation in Arctic governance. *Asian Countries and the Arctic Future,* 217–235.

[54] Kopra, S. (2019, August 22). *Will the Ice Silk Road become the compass for the Arctic?* Helsinki Institute of Sustainability Science.

[55] Sheng, L. (2014a). Capital controls and international development: A theoretical reconsideration. *Global Policy, 5*(1), 114–120.

[56] Sheng, L., & do Nascimento, D. F. (2021b). *Love and trade war: China and the US in historical context.* Springer.

to overthrow the political status quo in the Arctic or the existing world order. In its Arctic cooperation, China wants to build an image of a "near-Arctic state" that exercises "responsible power" and is an "important and legitimate stakeholder."[57] China strongly positions itself as merely an active participant, builder, and contributor to Arctic affairs. In a press briefing on the release of the Arctic policy white paper, Chinese Vice-Minister of Foreign Affairs Kong Xuanyou emphasized two positions that China will not adopt in its role as an Arctic stakeholder: it will not overstep and it will not be absent.[58] During this press conference, it was also stated that there would be a focus on protecting the interests of Arctic residents and indigenous groups and a commitment to maintaining peace and stability in the Arctic region, thereby effectively reducing the scruples of Russian authorities over cooperation with China in the Arctic region and strengthening the willingness of the two countries to cooperate. Moreover, the white paper on China's Arctic policy clearly points out that Arctic affairs are no longer just an issue between Arctic countries or within the Arctic region but are now a global issue affecting the overall interests of countries within and outside of the Arctic region, and the international community at large. It defines *respect* as the key basis, *cooperation* as an effective means, *win-win results* as the value being pursued, and *sustainability* as the fundamental goal of China's participation in Arctic affairs.[59]

There has been a historical breakthrough in cooperation between China and Russia in terms of both oil and gas cooperation, which has laid a good foundation for a win-win situation in the Arctic. Under China's oil trade agreement with Russia in 2013, it will purchase more than US$355 billion worth of crude oil from Russia in future; in exchange, Russian oil giant Rosneft will allow PetroChina to develop three offshore oil fields in the Sea of Okhotsk.[60] In 2014, when Russia and Western powers were quarreling over Ukraine and following Russia's annexation

---

[57] Mariia, K. (2019). China's Arctic policy: Present and future. *The Polar Journal, 9*(1), 94–112.

[58] Kong, X. (2018). SCIO briefing on China's policy on the Arctic (Translation provided by State Council Information office of the PRC).

[59] Xinhua. (2018, June 26). *China's Arctic policy.*

[60] China Petroleum News Center. (2021, January 5). *Nengyuan zhuanxing qushi xia eluosi de beiji nengyuan zhanlue* [Russia's Arctic energy strategy under the trend of energy transition].

of parts of Ukraine, the European Union pledged to reduce its dependence on Russian gas and applied sanctions that restricted the sources of funds and advanced technology for the Russian oil and gas industry. These kinds of sanctions not only limit the financing capabilities of Russian energy companies but also restrict Russian companies from obtaining cutting-edge technologies from Western companies through international cooperation. Moreover, these sanctions created great uncertainty in the Russian oil and gas sectors, which led Russia to subsequently seek cooperation in these sectors with emerging Asian economies like China. In May 2014, during Russian President Vladimir Putin's visit to China, the two countries signed the Sino–Russian East Route Natural Gas Cooperation Agreement, which marked the starting point of Russia's energy strategy to move eastward toward Asia. According to the agreement, Russia will supply China with 38 billion cubic meters of gas every year for 30 years through the China–Russia Eastern Gas Pipeline, which will benefit more than 500 million citizens in China.

However, Europe's dependence on Russian gas is also a strong reminder to China of the substantial harm it would suffer if Russia were to cut off its energy supplies. China is therefore compelled to seek active cooperation with Russia while ensuring it has a pathway back toward independence. To minimize energy security risks and reduce its dependence on Russian resources, China must diversify its investments and find additional energy suppliers and transportation routes while drawing closer to Russia. China has taken two steps in this direction. First, it has increased its investments in renewable energy. In 2020, China's overseas investment in solar photovoltaic, wind, and hydropower achieved a leading position, and its renewable energy investments accounted for 28% of its global investments.[61] Second, China is also actively seeking other partners and exploring new transportation routes. For example, the launch of the China–Myanmar gas pipeline and the construction of the China–Pakistan Economic Corridor will alleviate China's dependence on Russian energy imports.

---

[61] Secretariat, R. (2020). *Renewables 2020 global status report.* Rep. Paris: REN12.

6.4   Summary

Over billions of years, as the planet has shifted and the stars have moved, the vicissitudes of life and changes in the climate have promoted the evolution of species again and again. For the weak, change can lead to extinction, but for the strong, it can represent a new starting point. With global warming melting glaciers, Russia faces the challenges of permafrost degradation and reduced food production but also sees opportunities, especially with the opening of the Arctic shipping route, which has put the country in the world's spotlight. The Arctic climate issue is a double-edged sword that must be wielded wisely. International action on climate change has been through a long process of development and maturation, from questions over the conclusion that human activity causes climate change to the signing of the first climate convention to an increasing number of countries joining climate negotiations and devoting themselves to the development of renewable energy. Some countries have been able to fulfill their commitments and adhere to the carbon neutral concept from the beginning, and others have gone through erratic changes in their conception of the issue but eventually converged on the general trend. With China's participation, the growing cooperation between Russia and China in the Arctic has given Russia an additional bargaining chip in the Arctic game with the United States. This comes in the form of China's determination to participate in the development of the Arctic region, presenting a strong shield for Russia against American provocations, alarming the United States, and leading to the formation of a new triangular structure. Meanwhile, China does not neglect its need for a certain autonomy in Sino–Russian cooperation in fossil fuel energy resources, working to ensure its energy security and to avoid excessive dependence on any single country.

References

Arvin, J. (2021). *The latest consequence of climate change: The Arctic is now open for business year-round*. Vox. https://www.vox.com/22295520/climate-change-shipping-russia-china-arctic

Biden, J. (2021). *The Biden plan for the clean energy revolution and environmental justice*. https://joebiden.com/climate-plan/

Borger, J. (2001). Bush kills global warming treaty. *The Guardian*. https://www.theguardian.com/environment/2001/mar/29/globalwarming.usnews

British Petroleum. (2021). *Statistical review of world energy—2021 Russia's energy market in 2020.* https://www.bp.com/content/dam/bp/business-sites/en/global/corporate/pdfs/energy-economics/statistical-review/bp-stats-review-2021-russia-insights.pdf

Bush, G. W. (2001, March). *Text of a letter from the president to Senators Hagel, Helms, Craig, and Roberts.* White House Archives. https://georgewbush-whitehouse.archives.gov/news/releases/2001/03/20010314.html

Bush, G. W. (2021). *Clean energy and climate change.* Bush White House Archives. https://georgewbush-whitehouse.archives.gov/ceq/clean-energy.html

Champion, M., & Doff, N. (2021). *Russia's getting left behind in global dash for clean energy.* Bloomberg. https://www.bloomberg.com/news/articles/2021-03-15/russia-s-getting-left-behind-in-global-dash-for-clean-energy

Chen, G. (2012). China's emerging Arctic strategy. *The Polar Journal, 2*(2), 358–371.

Chenguang|Zhuo yan tianxia: Yongjiu dongtu ronghua quanqiu weiji jiaju [Chenguang|Focus on the world: Melting permafrost global crisis intensifies]. https://www.8world.com/stories/morning-express/spotlight-russia-permafrost-1590451

China Petroleum News Center. (2021). *Nengyuan zhuanxing qushi xia eluosi de beiji nengyuan zhanlue.* [Russia's Arctic energy strategy under the trend of energy transition]. http://news.cnpc.com.cn/system/2021/01/05/030021033.shtml

Christopher, T. (2017). *Trump signs executive order to roll back Obama-era climate actions, power plant emissions rule.* CNBC. https://www.cnbc.com/2017/03/27/trump-to-roll-back-obama-climate-actions-power-plant-emissions-rule.html

Climate Action Tracker. (2021). *Russian Federation.* https://climateactiontracker.org/countries/russian-federation/

CNN. (2002). *Bush unveils voluntary plan to reduce global warming.* https://edition.cnn.com/2002/ALLPOLITICS/02/14/bush.global.warming/index.html

Conley, H. A. (2021). *Climate change will reshape Russia.* Center for Strategic International Studies report. https://www.csis.org/analysis/climate-change-will-reshape-russia

FACT SHEET. (2021). *President Biden sets 2030 greenhouse gas pollution reduction target aimed at creating good-paying union jobs and securing U.S. leadership on clean energy technologies.* White House briefing. https://www.whitehouse.gov/briefing-room/statements-releases/2021/04/22/fact-sheet-president-biden-sets-2030-greenhouse-gas-pollution-reduction-target-aimed-at-creating-good-paying-union-jobs-and-securing-u-s-leadership-on-clean-energy-technologies/

Fang Lexian, & Wang Yujing. (2021). Oumeng yu eluosi qihou hezuo de jinzhan yu juxian [Progress of EU-Russian cooperation on climate]. *Heping yu Fazhan* [Peace and Development]. 2021(03).

GECF. (2021). *Global gas outlook 2050.* https://www.gecf.org/_resources/files/events/gecf-to-unveil-latest-edition-of-flagship-publication-global-gas-outlook-2050/2020-edition-of-the-gecf-global-gas-outlook-2050.pdf

Grantham Research Institute on Climate Change and the Environment. (2021). *Greenhouse gas emission reduction* (Presidential Decree 752). https://www.climate-laws.org/geographies/russia/policies/greenhouse-gas-emission-reduction-presidential-decree-752

Hovi, J., Sprinz, D. F., & Bang, G. (2010). Why the United States did not become a party to the Kyoto Protocol: German, Norwegian, and US perspectives. *European Journal of International Relations, 18*(1), 129–150.

Kelemen, D. R., & Vogel, D. (2010). Trading places: The role of the US and EU in international environmental politics. *Comparative Political Studies, 43*(4), 427–456.

Kiselev, S., Romashkin, R., Nelson, G. C., Mason-D'Croz, D., & Palazzo, A. (2013). Russia's food security and climate change: Looking into the future. *Economics, 7*(1), 1–66.

Kong, X. (2018). *SCIO briefing on China's policy on the Arctic.* State Council Information office of the PRC. http://www.scio.gov.cn/32618/Document/%201618357/1618357.htm

Kopra, S. (2019). *Will the Ice Silk Road become the compass for the Arctic?* Helsinki Institute of Sustainability Science report. https://www2.helsinki.fi/en/news/society-economy/will-the-ice-silk-road-become-the-compass-for-the-arctic

Liang, Y. (2019). *China, Russia agree to upgrade relations for new era.* Xinhua News Agency. http://www.xinhuanet.com/english/2019-06/06/c_138119879.htm

Lyddon, C. (2021). *Focus on Russia.* World Grain. https://www.world-grain.com/articles/14975-focus-on-russia

Mariia, K. (2019). China's Arctic policy: Present and future. *The Polar Journal, 9*(1), 94–112.

Milman, O. (2021). Biden returns US to Paris climate accord hours after becoming president. *The Guardian.* https://www.theguardian.com/environment/2021/jan/20/paris-climate-accord-joe-biden-returns-us

Moscow Times. (2021). *Skepticism to acceptance: How Putin's views on climate change evolved over the years.* https://www.themoscowtimes.com/2021/07/01/skepticism-to-acceptance-how-putins-views-on-climate-change-evolved-over-the-years-a74391

Obama, B. (2015, December 12). *Statement by the President on the Paris Climate Agreement*. White House Archives. https://obamawhitehouse.archives.gov/the-press-office/2015/12/12/statement-president-paris-climate-agreement

Oshchepkov, M. (2021). *Russia has set an ambitious goal for reducing emissions by 2030*. Climate Scorecard. https://www.climatescorecard.org/2021/07/russia-has-set-an-ambitious-goal-for-reducing-emissions-by-2030/

Outka, U. (2016). The Obama administration's Clean Air Act legacy and the UNFCCC. *Case Western Reserve Journal of International Law, 48*(1), 109–125.

Overseas Project Encyclopedia. *Dujia: Yamal LNG project: bingshang sichouzhilu de zhongyao jiedian* [Yamal LNG project: An important pivot point of the Ice Silk Road], China Belt and Road Network. https://www.sohu.com/a/450900494_731021

Parker, C. F., & Karlsson, C. (2018). The UN climate change negotiations and the role of the United States: Assessing American leadership from Copenhagen to Paris. *Environmental Politics, 27*(3), 519–540.

Pincus, R. (2020). Three-way power dynamics in the Arctic. *Strategic Studies Quarterly, 14*(1), 40–63.

REN21. (2020). *Renewables 2020 global status report*. https://ren21.net/gsr-2020/

Semenov, A. (2021). *Russian political forces meet climate change*. Center for Strategic International Studies report. https://www.csis.org/analysis/russian-political-forces-meet-climate-change.

Sharmina, M., Anderson, K., & Bows-Larkin, A. (2013). Climate change regional review: Russia. *Wiley Interdisciplinary Reviews: Climate Change, 4*(5), 373–396.

Sheng, L. (2014a). Capital controls and international development: A theoretical reconsideration. *Global Policy, 5*(1), 114–120.

Sheng, L. (2014b). Economic structure, cost outsourcing and global imbalances. *Journal of Australian Political Economy, 74*, 81–95.

Sheng, L. (2015). Theorizing income inequality in the face of financial globalization. *The Social Science Journal, 52*(3), 415–424.

Sheng, L. (2021). *How Covid-19 reshapes new world order: Political economy perspective*. Springer.

Sheng, L., & do Nascimento, D. F. (2021a). *The belt and road initiative in south-south cooperation: The impact on world trade and geopolitics*. Palgrave MacMillan.

Sheng, L., & do Nascimento, D. F. (2021b). *Love and trade war: China and the US in historical context*. Springer.

Swanson, C., & Levin, A. (2020). *Liquefied natural gas exports are a climate threat*. NRDC report. https://www.nrdc.org/experts/christina-swanson/liquefied-natural-gas-exports-are-climate-threat

Tynkkynen, V. P., & Tynkkynen, N. (2018). Climate denial revisited: (Re)contextualising Russian public discourse on climate change during Putin 2.0. *Europe-Asia Studies, 70*(7), 1103–1120.

Union of Concerned Scientists. (2020). *Each country's share of* CO_2 *emissions*, updated August 12, 2020. https://www.ucsusa.org/resources/each-countrys-share-co2-emissions

United Nations. (2021). *Kyoto Protocol.* https://unfccc.int/kyoto_protocol

US Senate. (1997). *Resolution 98.* https://www.congress.gov/bill/105th-congress/senate-resolution/98/committees

Vyalov, S. S., Gerasimov, A. S., & Fotiev, S. M. (1998). Influence of global warming on the state and geotechnical properties of permafrost. *Permafrost: Proceedings of the Seventh International Conference*, 1097–1102.

Wishnick, E. (2021). Will Russia put China's Arctic ambitions on ice? *The Diplomat.* https://thediplomat.com/2021/06/will-russia-put-chinas-arctic-ambitions-on-ice/

Xinhua News Agency. (2018). *China's Arctic policy.* http://english.www.gov.cn/archive/white_paper/2018/01/26/content_281476026660336.htm

Yin, Y. C., & Sheng, L. (2021). Theorizing about global imbalances An inequality perspective. *Argumenta Oeconomica, 46*, 169–181.

Zhang, P., & Yang, J. (2016). Changes in the Arctic and China's participation in Arctic governance. *Asian Countries and the Arctic Future*, 217–235. https://doi.org/10.1142/9789814644181_0014

Extra-Regional Players in the Arctic: EU, China, Japan, Singapore, and South Korea

Abstract The melting of Arctic ice caps in recent decades has added to the complexity of competition in the region, as it opens the door for the involvement of extra-regional players, such as China, Japan, Singapore, South Korea, and the European Union. China is strengthening its influence in the Arctic by enhancing its bilateral and multilateral cooperation in the region. The participation of China and other new players in discussions on Arctic affairs is changing the balance of power in the region.

Keywords Arctic · European Union · Emerging powers · Asian countries

The Arctic is one of the most resource-abundant regions in the world. It has been nearly impossible for human beings to explore the Artic in the past millennium, owing to technological limitations and the region's harsh environmental conditions. Although technological advancements in the last two centuries have made developments in the Arctic possible, these opportunities have remained the privilege of a small pool of countries that include the United States, Russia, Canada, and a few countries in Northern Europe. In recent decades, human activity in the Arctic has largely been limited to an exclusive "club" comprising the countries in

E. L. Sheng, *Arctic Opportunities and Challenges*,
https://doi.org/10.1007/978-981-19-1246-7_7

its periphery, and this club has essentially been dominated by the United States and Russia.

Increasing global warming in recent decades has fundamentally changed the game in the Arctic. On the one hand, warmer temperatures have not only led to the melting of the polar ice caps and changes in the Arctic landscape but also reduced the costs of exploiting natural resources. For instance, the melting of ice caps has facilitated the development of several new routes in the region, which have created a range of opportunities for polar tourism and improved transport logistics. On the other hand, the melting of ice caps in the Arctic also leave the door open to extra-regional players, such as China, Japan, Korea, and the European Union. The power balance in the region has changed especially following China's involvement. That is, the bipolar system that was previously dominated by Russia and the United States has been replaced by a new structure, which some have termed the "Strategic Triangle"[1,2] and described as a "zero-sum game."[3] Consequently, regional issues relating to the Arctic have expanded into more complicated and complex international issues. More intense competition has unfolded as the countries involved have sought to maximize their own economic and political benefits.

Although the three giant powers, China, Russia, and the United States, are playing leading roles in the game, one cannot neglect the roles of intermediary players such as Japan and South Korea. Despite being geographically distant from the region and lacking legitimacy for their involvement, these two extra-regional players have a deep interest in the Arctic. The Arctic's wealth of minerals and marine resources is of great value to both countries as they are resource deficient. In addition, the two countries stand to benefit from the development of polar routes in northeastern Asia.

[1] Huebert, R. (2019). The New Arctic Strategic Triangle Environment (NASTE). In *Breaking the ice curtain? Russia, Canada, and Arctic security in a changing circumpolar world* (pp. 75–93).

[2] Pincus, R. (2020). Three-way power dynamics in the Arctic. *Strategic Studies Quarterly, 14*(1), 40–63.

[3] Wang, X. Q. (2020). An analysis of the law of China-US-Russia Triangle Relations. *Studies in Russia, Eastern Europe and Central Asia, 2020*(3), 32–52.

In this chapter, we discuss the roles of extra-regional players in the Arctic game. These include China, Japan, South Korea, and the European Union. We argue that China has the strongest influence on Arctic affairs owing to its rising capabilities in international politics, the global economy, and polar technology. In this regard, it is necessary to understand how China interacts with other extra-regional players outside the Arctic.

7.1 THE EUROPEAN UNION: A SEMI-REGIONAL PLAYER

The European Union is strictly a "semi-peripheral" Arctic entity. Of its 27 members states, only 3—Denmark, Finland, and Sweden—are located near the Arctic. These three states are permanent member states of the Arctic Council, a leading intergovernmental forum that aims to promote cooperation in the Arctic.[4] Aside from the three states, a few other member states of the European Union are either directly or indirectly involved in Arctic issues.

7.1.1 The European Union's Arctic Policy

In 2007, recognizing the impact that climate change in the Arctic would have on energy and shipping in Europe, the European Parliament urged the European Union to play an active role in the Arctic region. These efforts eventually culminated in the European Commission finally adopting a communication titled "The EU and the Arctic Region" in November 2008. This document sets out three policy objectives for Arctic governance: working with the local people and indigenous communities in the Arctic region to protect the Arctic, promoting the sustainable use of resources, and contributing to the strengthening of multilateral governance in the Arctic.[5] In 2009, the European Union also signed agreements for economic and geological cooperation with two Arctic countries, Norway and Iceland. As the European Union's primary

---

[4] Arctic Council. The Arctic Council.

[5] Perez, E. C., & Yaneva, Z. V. (2016). The European Arctic policy in progress. *Polar Science, 10*(3), 441–449.

partner in the Arctic, the Norwagian government frequently encourages European leaders to expand the European Union's influence in this region.[6]

The European Union has recently enhanced its funding and cooperation with Arctic countries under the "EU Strategy for the Baltic Sea Region."[7] The strategy includes three primary objectives, "save the sea, connect the region and increase prosperity."[8] Unlike the superpowers, whose interest lies mainly in energy extraction, the European Union focuses on creating long-term sustainable economic development, tackling climate change problems, and facilitating dialogue between Arctic countries.[9] For instance, in 2008, it focused on environmental and security issues relating to resource competition or military presence in the Arctic. Nonetheless, as soon as peace was restored to the region, the European Union no longer had any incentive to be actively involved in Arctic issues.[10] It was only in 2016, when the European External Action Services Commission Joint Communication titled "An integrated European Union Policy for the Arctic" was introduced, that a greater emphasis was placed on the European Union's response to special challenges relating to Arctic issues. Ever since, the European Union has actively promoted cooperative research efforts in the Arctic region. Overall, the focus of its Arctic policy has shifted in response to previous concerns about politics and security to the gathering of soft power built on science and research.[11] At present, the European Union's Arctic policy remains moderate and progressive.

[6] Offerdal, K. (2011). The EU in the Arctic: In pursuit of legitimacy and influence. *International Journal,* 66(4), 861–877.

[7] Stępień, A., & Koivurova, T. (2017). *Arctic Europe: bringing together the EU Arctic.* Policy and Nordic cooperation.

[8] *EU strategy for the Baltic Sea region.* (2018). Publications Office.

[9] Martynova, M. (2019). Eu, Russia and China Arctic Strategies: Comparative analysis. *Economic and Social Development: Book of Proceedings,* 773–778.

[10] Østhagen, A. (2019). *The New geopolitics of the Arctic: Russia, China, and the EU.* Wilfried Martens Centre for European Studies.

[11] Perez, E. C., & Yaneva, Z. V. (2016). The European Arctic policy in progress. *Polar Science,* 10(3), 441–449.

7.1.2   Cooperation Between China and the European Union in the Arctic

As both China and the European Union share mutual economic and scientific interests in the Arctic, they are incentivized to cooperate. The European Union strongly values China's extensive experience in conducting polar scientific research as well as its advanced capabilities in this area.[12] For instance, its first icebreaker, "Xue Long," was invented in 1993, and "Xue Long 2" was introduced and deployed for scientific research in 2019.[13] Moreover, both China and the European Union support free shipping through the Arctic and mutually agree to establish new logistics routes.[14] The planned Polar Silk Road will not only provide new opportunities for cooperation between China and the European Union but will also catalyze further joint developments between Asia and Europe.

Nonetheless, the European Union still has major concerns about China's participation in Arctic affairs. Pelaudeix[15] argued that China's involvement in the Arctic represents a threat to the security of the European Union as well as that of the polar countries. First, as a dominant financial power, China may directly or indirectly control Arctic states with poorly developed economies, such as Greenland. Large Chinese state-owned enterprises may potentially exclude domestic Arctic enterprises, threatening their survival. Second, military cooperation between Russia and China threatens the security of the European Union. While China does not maintain military bases or airports in the Arctic and in Europe, Russia may share its military infrastructure with China, allowing the latter to use these facilities to project its military power and to further expand its influence over Europe. Finally, China's extraction of natural resources in the Arctic may create severe pollution and other environmental problems in the region.

[12] Jakobson, L. (2010). China prepares for an ice-free Arctic. *SIPRI Insights on Peace and Security, 2010*(2).

[13] Xia, H. (Ed.). (2019, November 25). *China's polar icebreaker Xuelong 2 finished its icebreaking tasks.* XinhuaNet.

[14] Daemers, J. (2012). *The European Union in the Arctic: A Pole Position?* Bruges Regional Integration & Global Governance.

[15] Pelaudeix, C. (2018). *Along the road: China in the Arctic* (European Union Institute for Security Studies Briefs 13).

Although China and Europe have a common interest on some issues in the Arctic region, their diverging interests on others have been an obstacle to further collaboration.[16] The situation is further complicated by the European Union's lack of a unified policy attitude toward the Arctic, because its member states include both Arctic and non-Arctic countries, which have different interests in the region. These diverse voices and opinions often result in the adoption of inconsistent attitudes toward Arctic affairs.[17] To achieve a more effective Arctic policy, the European Union needs to raise greater awareness of the problems facing the Arctic, so that its member states and their respective leaders can better grasp the relevance of the European Union to the Arctic and the significance of the Arctic to non-Arctic countries. The European Union must also adopt a long-term strategy toward the Arctic because it is an area of increasing geopolitical importance. If it plans to use this new platform to expand its geopolitical influence in the Arctic, it must not only consider its relationship with China, which is becoming increasingly powerful, but also Russia and its deteriorating bilateral relations with member states of the European Union, especially after the conflict and confrontation with Russia over the Ukrainian and Crimean crises. The European Union also needs to keep up with the United States' latest policy developments toward the Arctic, given that it is another major player in the region.

7.2   PLAYERS FROM EAST ASIA: JAPAN, SOUTH KOREA, SINGAPORE, AND CHINA

In the last decade, Arctic affairs have not only garnered the attention of countries in the region but also aroused the interest of the wider international community. This has led to multiple non-Arctic countries applying to become observers in the Arctic Council. In May 2013, the applications of China, Japan, Singapore, and South Korea were accepted during the ministerial conference of the Arctic Council. This was a watershed moment for the representation of East Asian countries in Arctic affairs. As observers in the Arctic Council, these East Asian states would be better assured of operating in a secure environment provided by the Arctic

---

[16] Sheng, L. (2021). *How Covid-19 reshapes new world order: Political economy perspective.* Springer.

[17] Østhagen, A. (2019). *The new geopolitics of the Arctic: Russia, China, and the EU.* Wilfried Martens Centre for European Studies.

states. In addition, several East Asian states have an economic interest in the Arctic. For instance, South Korea, which is keenly aware of the lack of immediate commercially viable options for Arctic shipping,[18] remains interested in the construction of new shipping routes in the Arctic, as these will benefit its shipbuilding, shipping, and energy sectors.

Despite being non-Arctic countries with little territorial claim over the region, the growing interest of East Asian countries in Arctic affairs mainly stems from their concerns about climate change and geopolitical rivalries. At the same time, the interest of some of these countries is defined under international legal regimes (discussed in the previous section), such as the right of innocent passage in the NSR, the right to the common heritage of mankind in the seabed of the Arctic Ocean, and the right to conduct scientific research. In this context, East Asian countries have been more vocal in expressing their interests and concerns over affairs regarding the Arctic.

The burgeoning interest of East Asian countries in the Arctic mainly pertains to four areas. The first area is environmental concern. As Babin and Lasserre noted, "if the coastal states are directly impacted by climate change in the Arctic Ocean, the major Asian powers will eventually also be affected as several climate models show a link between Arctic climate conditions and prevailing conditions in Eastern Asia."[19] The second area pertains to natural resources, which many economically powerful countries in East Asia are lacking. These shortages in natural resources motivate East Asian countries to advocate for better conservation of Arctic resources. In fact, countries such as South Korea and Japan already expressed their interest in this issue in 2009. Nonetheless, as they do not possess any territory in the Arctic, the interest of these East Asian players has largely been overshadowed by that of Arctic states.[20] The third area pertains to economic interests. The main concern of export-oriented economies in East Asia is that the exploitation of Arctic resources and the NSR may fundamentally change the landscape of the world economy.

[18] Solli, P. E., Rowe, E. W., & Lindgren, W. Y. (2013). Coming into the cold: Asia's Arctic interests. *Polar Geography, 36*(4), 253–270.

[19] Babin, J., & Lasserre, F. (2019). Asian states at the Arctic Council: Perceptions in Western states. *Polar Geography, 42*(3), 145–159.

[20] Ibid.

In their comparative studies of the Arctic strategies of East Asian countries, Solli, Rowe, and Lindgren[21] argued that East Asian countries are mainly concerned with economic opportunities and environmental issues that pertain to the Arctic and less with security. In this section, we present brief case studies of East Asian countries that are extra-regional players in the Arctic: China, Japan, Singapore, and South Korea. The fourth area pertains to scientific research. As Sakhuja and Narula noted, "scientific research is a common area of interest for all Asian countries in the Arctic. Many of these countries including China have set up polar research stations in the Arctic. Japan was the first Asian country to undertake Arctic scientific research and to determine the viability of the NSR. The Republic of Korea has the advantage of a well-developed ship building industry to provide ships capable of traversing through the ice. Singapore has a technological edge in marine industries, particularly in management of ports, deployment of offshore marine and engineering, and has a keen interest in the development of international maritime policy."[22]

7.2.1   China as an Emerging Power in the Arctic

To China, the Arctic represents important environmental concerns in terms of both natural resources and climate change and an area for economic opportunities. As China's representative to the Arctic stated, "China's Arctic interest is grounded in China's northerly geographical location, concern about global climate change and interest in the potential impact on China of new Arctic shipping routes affecting energy and trade."[23] This explains China's eagerness to be accepted as an observer in the Arctic Council and to promote the transparency and credibility of this institution by strengthening its presence. Nonetheless, it should be noted that the Arctic is not a top priority in China's foreign policy concerning its economic opportunities.[24]

---

[21] Solli, P. E., Rowe, E. W., & Lindgren, W. Y. (2013). Coming into the cold: Asia's Arctic interests. *Polar Geography, 36*(4), 253–270.

[22] Sakhuja, V., & Narula, K. (Eds.). (2016). *Asia and the Arctic: Narratives, perspectives and policies.* Springer.

[23] Zhao, J. (2013, January 21). Speech by Ambassador Zhao Jun at the Arctic Frontiers Conference.

[24] Solli, P. E., Rowe, E. W., & Lindgren, W. Y. (2013). Coming into the cold: Asia's Arctic interests. *Polar Geography, 36*(4), 253–270.

By dint of the successful export-oriented strategy, China have enjoyed impressive economic growth over the past few decades.[25],[26] As a result, China has strengthened its influence in the Arctic by enhancing its bilateral and multilateral cooperation in the region in recent years. In 2014, China's President Xi proposed the Silk Road Economic Belt and the 21st Century Maritime Silk Road under the BRI. The implementation of the BRI is aimed at further enhancing China's economic openness and attracting more foreign direct investment.[27],[28] This also represented a major avenue for China to participate in the international stage and outlined the new direction of its foreign policy strategy.[29],[30] Subsequently, in 2017, China and Russia jointly proposed the Polar Silk Road to facilitate additional development and cooperation in the Arctic region.[31] In the following year, the State Council Information Office of the People's Republic of China issued its Arctic policy white paper, outlining the objective, position, principle, and character of China in the region. At the end of 2019, China officially announced that a new satellite would be launched in 2022 to "track shipping routes and monitor changes in sea ice in the Arctic." Furthermore, China's latest Five-Year Plan for the 2021–2025 period states that the country is achieving its goal of constructing the Polar Silk Road and is making concerted efforts to participate actively in Arctic affairs. Overall, these progressive advancements by China in the Arctic region demonstrate the country's interest and determination in pursuing development in multiple areas pertaining to the Arctic.

[25] Sheng, L. (2014b). Economic structure, cost outsourcing and global imbalances. *Journal of Australian Political Economy, 74*, 81–94.

[26] Sheng, L. (2015). Theorizing income inequality in the face of financial globalization. *The Social Science Journal, 52*(3), 415–424.

[27] Gu, X., & Sheng, L. (2010). A sensible policy tool for Pareto improvement: capital controls. *Journal of World Trade, 44*(3).

[28] Sheng, L. (2014a). Capital controls and international development: A theoretical reconsideration. *Global Policy, 5*(1), 114–120.

[29] Sheng, L. & do Nascimento, D. F. (2021a). *The Belt and Road Initiative in South-South Cooperation: The impact on world trade and geopolitics.* Palgrave Macmillan.

[30] Sheng, L., & Zhang, Z. Y. (In press). *Revisiting global imbalances: A comparative analysis of income and consumption inequality.* Economics & Politics.

[31] Yao, Z. (2019, April 7). Ice Silk Road framework welcomed by countries, sets new direction for Arctic cooperation.

As one of the largest consumers of energy in the world, China is relatively vulnerable to supply changes and price fluctuations in the global energy market. China's heavy dependence on external energy supplies also makes the country sensitive to developments in its maritime trade routes, especially those in the south. China's energy is mainly sourced from the Middle East, Russia, South Asia, and South America.[32] For these reasons, and given its rapid economic development, energy security is a priority for China. However, the country faces insecurity in this area given its intensifying competition with the United States.

In its efforts to secure its energy supply, China has focused heavily on environmental research, shipping, and energy investments in the Arctic region.[33] Separately, the growing influence of geopolitics and maritime security issues has strengthened the political reasons for China to cooperate with Russia for its energy supply. First, energy trade between China and Russia can remain stable over a long period owing to the land border shared by the two countries, which make methods for supplying energy to China relatively simple. For instance, transportation can be used, as this is far less constrained by natural conditions than land pipelines. Second, the domestic conditions of the two sides are stable under their strong leadership, which promotes consistency in their internal and external policies. Third, as the countries are strategic partners—a result of their decades-long stable bilateral relations—the transportation of energy between Russia and China over land is seldom affected by natural factors and issues of national security and politics.

Unlike its energy trade with Russia, China's energy trades with the Middle East, South America, and South Asia are not only vulnerable to unstable natural conditions in the seas but also to international political factors, such as domestic turmoil in these countries and the deterioration of China's relations with the countries, which can lead to maritime disputes and even conflict. Given this uncertainty and instability in its imports of energy, China has sought to explore new sources of energy. Energy resources in the Arctic could help China increase its capacity for risk management while simultaneously reducing the cost of China's consumption of and trade in energy. This will help address

---

[32] International Trade Administration (ITA). (2021, February 4). *China's energy data*.

[33] Campbell, C. (2013). *China and the Arctic: objectives and obstacles*. US-China. Economic and Security Review Commission.

China's structurally induced consumption inequality and regional development disparities.[34] However, China's resource development in the Arctic has attracted the attention of other countries, especially the Arctic states. Moreover, China's plans to expand its energy imports—such as by opening channels to further promote economic development—have been seen by many countries in the wider international political arena as another ambitious means to strengthen its regional and international influence. As Oksana Antonenko, Director at the Global Political Risk team at the United Kingdom-based Control Risk consultancy, noted, the Arctic states view China as a near-Arctic state that is "playing a much more assertive role unilaterally."[35]

China also intends to achieve several economic and political objectives. From an economic perspective, China aims to gain new momentum in its economic development and to acquire advanced technologies and industrial capacities to secure its energy supply.[36] From a political perspective, China aims to use the Arctic as a platform to enhance its international influence and to integrate the BIR and SIR to promote more comprehensive foreign strategies.

However, China currently has insufficient economic, political, and technological capacity to dominate the Arctic for three main reasons. First, China is a non-Arctic country. Second, it has insufficient technological expertise in the extraction of natural minerals to pursue this independently in the Arctic. Last, Arctic countries have limited China's access to the region. As a consequence, China has sought to cooperate with one of the strongest states in the Arctic region, Russia. This cooperation could allow China to overcome the barriers to access imposed by Arctic countries and simultaneously acquire the technological expertise for natural resource extraction (i.e., from Russia). In addition, Bai and Voroenko suggested that "cooperation with Russia will give Chinese actions in the region more validity."[37]

---

[34] Yin, Y. C., & Sheng, L. (2021). Theorizing about global imbalances: An inequality perspective. *Argumenta Oeconomica, 46,* 169–181.

[35] Ng, A. (2021, May 22). *Tensions likely to grow as China seeks a bigger role in the Arctic.* CNBC.

[36] Pincus, R. (2020). Three-way power dynamics in the Arctic. *Strategic Studies Quarterly, 14*(1), 40–63.

[37] Jiayu, B. A. I., & Voronenko, A. (2016). Lessons and prospects of Sino-Russian Arctic cooperation. *Advances in Polar Science* (3).

Although China and Russia share a broad common interest that may form the basis for their cooperation in Arctic affairs, the two countries have different interests in other areas, which have led to clashes. Many of these can be traced to their conflicting strategic plans. For instance, as the largest country in the Arctic, Russia strives to maintain its dominance in the region's affairs and geopolitics; it maintains an attitude of exclusivity in Arctic affairs, seeking to minimize the participation of foreign countries. Yet, as a non-Arctic country, China strives to participate in Arctic affairs by pursuing bilateral and multilateral cooperation with Arctic countries and international organizations. In addition, because of historical constraints, Russia remains vigilant and concerned about China, and thus constantly has reservations about cooperation. Collectively, these differences in the interests of China and Russia are likely to create obstacles to the advancement of the Polar Silk Road.

Global warming has changed the circumstance in the Arctic landscape and deeply impacted the all countries in the world. Nonetheless, it has also created economic opportunities for many countries such as China, Japan, and South Korea, although these countries are external to the Arctic and regarded as "outsiders" to the region. While the status of "outsider" may have afforded little legitimacy to these countries in Arctic affairs, it has been insufficient to prevent their involvement entirely. Indeed, the three countries joined the Arctic Council as observer states in 2013.

In general, the three East Asian countries have similar interests and objectives in the North Pole. They are mainly concerned with diversifying their sources of energy, extracting natural minerals, developing climate change solutions, and building new routes for shipping and trading. However, because of their different strategies and state capacities, China, Japan, and South Korea tend to pursue different approaches to policymaking and engagement in the Arctic.

7.2.2   *Japan: A Neutral State in the Arctic*

As a self-identified maritime state, Japan has always expressed a willingness to contribute to scientific cooperation in the Arctic. Indeed, as one of the original 14 contracting parties of the original Spitsbergen Treaty in 1920, Japan has had a longer history of participation in Arctic affairs than the other East Asian states discussed in this chapter. Japan has also been actively engaged in the Arctic for over half a century.[38] In the 1950s, it began to conduct its first scientific observations and research projects in the Arctic. These initiated a longstanding tradition of Japanese scientific and academic work in the field of Arctic studies, which eventually secured the country's status as the first "regional outsider" in the Arctic Circle and saw its entry into the International Arctic Science Committee.[39] In 1991, Japan established an observation station in the Arctic.[40]

For the above reasons, Japan has accumulated a wealth of experience in conducting scientific research in the Arctic and has actively invested in the region's infrastructure. Nonetheless, these efforts have mainly been driven by economic motives comparing to other neighboring countries such as China. By 2010, China had replaced Japan as the world's most successful exporter and the second-largest economy.[41] Although shipping routes from the Arctic to Japan remain unprofitable, the Japanese government has retained its interest in this region because these routes are important for the country's energy security; they facilitate the transport of liquefied natural gas from Norway and Russia to Japan. Aside from the issue of energy security, Japan's strategy toward the Arctic has few concerns that pertain to other areas of security.[42] As Tonami Aki, a professor of international relations and economics at the University of Tsukuba, stated, "the changes in the Arctic do not pose direct security threats to Japan."[43]

Over the last decade, the Japanese government has continued to pursue its policies and strategies to develop the North Pole. In 2012, two major Japanese companies—Japan Oil, Gas, and Metals National Corporation

[38] Holroyd, C. (2020). East Asia (Japan, South Korea and China) and the Arctic. In *The Palgrave handbook of Arctic policy and politics* (pp. 319–332). Palgrave Macmillan.

[39] Almazova-Ilyina, A. B., Vinogradov, A. D., Krasnozhenova, E. E., & Eidemiller, K. Y. (2020). National interests of Japan and its emerging Arctic policy. In A. Seymour, *IOP Conference Series: Earth and Environmental Science* (Vol. 539, No. 1, p. 012048). IOP Publishing.

[40] Ibid.

[41] Sheng, L., & do Nascimento, D. F. (2021b). *Love and trade war: China and the US in historical context*. Springer.

[42] Ng, A. (2021, May 22). *Tensions likely to grow as China seeks a bigger role in the Arctic*. CNBC.

[43] Tonami, A. (2014). The Arctic policy of China and Japan: Multi-layered economic and strategic motivations. *Polar Journal, 4*(1), 105–126.

(JOGMEC) and International Petroleum Development Dishi Holdings (INPEX Holdings)—jointly established a new company, Greenland Petroleum Development, which won a license to develop the KANUMAS project as a regional seismic program in the icefilled arctic waters offshore North-West, East, and North-East Greenland. This permit allowed Japan to conduct vein and geological surveys in western Greenland. Greenland Petroleum Development also established joint ventures with oil companies in the United Kingdom, the United States, Norway, and Greenland to jointly develop Greenland's mineral resources. Japan has also launched multiparty cooperation with Russia in the fields of integrated energy and transportation, and has emphasized the strengthening of its bilateral relations with this major player in the Arctic.[44] In 2015, the Japanese government issued its Arctic policy, which outlined Japan's diverse interests and perspectives concerning the Arctic.[45] The specific initiatives under this policy were research and development, international cooperation and the sustainable use of natural resources.[46] Careful examination of this document reveals that the Japanese government has acknowledged the Arctic issue and that it regards climate change, shipping and enhancing the relations between Japan and Arctic countries as its long-term objectives in the region, thus acknowledging the importance of Arctic affairs to Japan. Nonetheless, the document fails to reveal the government's specific policy arrangements.[47]

In a policy brief delivered to Perpetual Industries Inc. PRPI directors by Kazuko Shiraishi at the 2016 Arctic Circle Assembly, the Japanese government specifically mentioned that it deemed climate change and sustainable development in the Arctic to be key national issues.[48] As a member of the international community, Japan considers itself to have an

[44] Ohnishi, F. (2016). Japan's Arctic policy development: From engagement to a strategy. In *Asian Countries and the Arctic Future*, 171–182.

[45] Japan's Arctic Policy. (2015). The Headquarters for Ocean Policy.

[46] Tonami, B. (2021). UK–Japan responses in the Arctic.

[47] Almazova-Ilyina, A. B., Vinogradov, A. D., Krasnozhenova, E. E., & Eidemiller, K. Y. (2020, July). National interests of Japan and its emerging Arctic policy. In *IOP Conference Series: Earth and Environmental Science* (Vol. 539, No. 1, p. 012048). IOP Publishing.

[48] The Polar Connection. (2021). *What is Japan's Arctic interest?*

obligation to preserve the natural environment.[49] Moreover, the dramatic melting of ice in the Arctic presents a threat to Japan. As an island country, Japan risks flooding with rising sea levels. A rise in sea levels would not only increase the costs of emergency disaster management and the construction of protective infrastructure but also directly threaten the lives of Japanese people. This serious threat has motivated Japan to cooperate with both Arctic and non-Arctic countries to reduce the severity of ice melt in the Arctic. Another motivating factor for Japan's involvement in these issues is the abundance of marine species in the Arctic, many of which are consumed by Japanese people each year. Access to the Arctic can allow Japan to enhance its food security and reduce the domestic prices of marine products. Aside from economic growth, Japan has prioritized food security and climate change as areas of state interest. The Japanese government allocated approximately US$13 million to matters relating to the Arctic in 2020. Much of this was spent on research and development; it included US$9.2 million dedicated to the launch of the ArCS II (the Arctic Challenge for Sustainability II) Project 57, a research project in which the majority of Japanese-based Arctic researchers participated.

Overall, Japan's interest in the Arctic is similar to that of China and other non-Arctic countries. From the perspective of Japan, the Arctic region contains economically valuable natural resources that are worth exploiting, as well as important solutions for addressing global climate change. Japan's special ambassador to the Arctic, Kazuko Shiraishi, confirmed this. However, these goals are often at odds with one another, and it remains unclear how a balance will be struck between these conflicting pursuits.[50]

7.2.3   South Korea in the Arctic

Like Japan, South Korea has been actively involved in Arctic affairs for decades, especially in the areas of polar research. Rather than security or political concerns, South Korea's interest in the Arctic has mainly pertained to the climate change and its economy, and the country has

---

[49] Tonami, A. (2014). The Arctic policy of China and Japan: Multi-layered economic and strategic motivations. *The Polar Journal, 4*(1), 105–126.

[50] The Polar Connection. *What is Japan's Arctic interest?* Retrieved September 20, 2021, from https://polarconnection.org/japan-arctic-interest/.

engaged in extensive polar research in the pursuit of these interests. On the topic of Arctic diplomacy, South Korea's representative to the region stressed the importance of addressing environmental challenges in the Arctic, emphasizing that climate change was a "global drama which calls for global efforts" and was an issue that concerned South Koreans as "global citizens."[51] According to the senior research fellow at the Korea Institute for Maritime Strategy. Sukjoon Yoon,[52] the dramatic changes in the Arctic landscape and natural environment driven by climate change had "led many maritime pundits in South Korea to propose that an 'Arctic Bonanza.'" Yoon articulated that the Arctic's rich natural resources could provide economic opportunities for South Korea, a country that was at risk of economic stagnation, because the Arctic could help to "boost South Korea's flourishing shipbuilding industry, supply plenty of cheap energy and rare mineral resources, and most importantly, create jobs for the new generation of young South Koreans."

Three main goals are reflected in South Korea's strategy toward the Arctic: to build partnerships with members of the Arctic Council, to advance global research in environmental science and engineering in the Arctic, and to secure stable energy supplies through the creation of new business opportunities in the Arctic and the development of bilateral economic relationships with Arctic nations.[53]

Much like other developed countries, South Korea has always been interested in the Arctic. In 1996, it established a joint Arctic research program with Japan; this subsequently became an independent program in 2001.[54] Published in 2013, the Korean Arctic Master Plan was a comprehensive policy for South Korean engagements in the Arctic, outlining the country's perspectives on business, scientific research, and international corporation in the region.[55]

---

[51] Lee, B. (2013, January 21). *Korea's Arctic policy: A Korean route towards the Arctic frontier*. Speech at the Arctic Frontiers Conference.

[52] Yoon, S. (2016). A cooperative maritime capacity-sharing strategy for the Arctic region: The South Korean perspective. In V. Sakhuja & K. Narula, *Asia and the Arctic* (pp. 49–62). Springer.

[53] Ibid.

[54] Holroyd, C. (2020). East Asia (Japan, South Korea and China) and the Arctic. In *The Palgrave handbook of Arctic policy and politics* (pp. 319–332). Palgrave Macmillan.

[55] Jin, D., Seo, W. S., & Lee, S. (2017). Arctic policy of the Republic of Korea. *Ocean & Coastal LJ, 22*, 85.

After obtaining observer status at the Arctic Council, South Korea also declared that it would make more visible commitment in the region, such as through the establishment of a long-term partnership with Arctic countries and an increase its scientific research activities.[56] Apart from advancing its scientific research, South Korea is motivated to develop new shipping routes in the Arctic, which will allow it to benefit from fossil fuel extraction. Like Japan, South Korea is highly dependent on fossil fuel imports from the Middle East and East Asia.[57]

In 2013, South Korea's Ministry of Foreign Affairs released a long-term policy report outlining five commitments to the Arctic region.[58] First, South Korea will respect international law and expect other countries to do the same. This emphasizes that the Arctic does not belong to any state and that Arctic security therefore represents an international issue. Second, South Korea supports the prevailing international maritime policies in the Arctic and will actively work to maintain the region's stable and conflict-free status. Third, South Korea anticipates a future in which the Arctic remains stable and conflict-free, and therefore, prioritizes approaches that will minimize risks to the Arctic. Fourth, South Korea will continue to make important contributions to the Arctic through its maritime capacity-sharing measures, as part of a regional commitment to Arctic security. Finally, by expanding its active bilateral cooperation with members of the Arctic Council, South Korea will make every effort to resolve outstanding practical problems.

Recent political instability in the Middle East has heightened South Korea's concerns about energy security. As a result, the extraction of natural resources in the Arctic represents an alternative and low-risk energy source for the country. Furthermore, the melting of ice in the Arctic may create new shipping routes that reduce the cost and time of trade voyage. Not only will this expand the trading volumes of South

---

[56] Krasnozhenova, E. E., Kulik, S. V., Chistalyova, T., Eidemiller, K. Y., & Karabushenko, P. L. (2021). The Arctic policy of the Republic of Korea. In *IOP Conference Series: Earth and Environmental Science* (Vol. 625, No. 1, p. 012011). IOP Publishing.

[57] Park, Y. K. (2014). South Korea's interests in the Arctic. *Asia Policy* (18), 59–65.

[58] Yoon, S. (2016). A cooperative maritime capacity-sharing strategy for the Arctic region: The South Korean perspective. In V. Sakhuja & K. Narula, *Asia and the Arctic* (pp. 49–62). Springer.

Korea's logistics companies but the country will also benefit from new trading paths to the European market.[59]

While attending the Third Eastern Economic Forum (EEF) in September 2017, South Korea's President Moon Jae-in presented a comprehensive account of the country's New Northern Policy to the international community. Also known as the "Nine Bridges Strategy," this policy focuses on cooperation with Russia in nine areas: Arctic shipping routes, natural gas, railways, ports, power, shipbuilding, agriculture, aquaculture, and industrial parks.[60] The policy is consistent with China's Polar Silk Road initiative, as both are northern-oriented development strategies and have overlapping objectives to promote economic integration in northern Eurasia. Theoretically, China and South Korea could complement each other's strengths in the economic sphere and cooperate in the fields of energy investments and financing.[61] However, South Korea's New North Policy is largely centered on the country's interests. Moreover, South Korea government fears that the country would be economically controlled by China if it were to engage in such cooperation. In addition, the creation of an Arctic sea route would mean that trade for both countries would have to pass through Japanese territorial waters, as the Sea of Japan is the gateway through which ships from China and South Korea enter and exit the Arctic waterway. Yet, Japan has historically experienced constant friction with these two Northeast Asian countries while adopting an attitude of "generous diplomacy" toward the United States. Thus, the attitude of Japan and the United States will determine the future directions of relations between China and South Korea in issues pertaining to the Arctic.

7.2.4   Singapore: A Tropical Player in the Arctic

Located at a strategic point along the Straits of Malacca, Singapore has played a key role in facilitating a shipping lane between Asia and Europe and consequently maintains a significant influence on the geoeconomics

---

[59] Holroyd, C. (2020). East Asia (Japan, South Korea, and China) and the Arctic. In *The Palgrave handbook of Arctic policy and politics* (pp. 319–332). Palgrave Macmillan.

[60] Zhang, D. M., Zhao, S. Y., & Zhang, D. Z. (2020). An analysis of the "New Northern Policy" of South Korea's Moon Jae-in Government. *Journal of Liaoning University* (Philosophy and Social Sciences Edition).

[61] Sheng, L. (2014c). The effects of foreign expansion on local growth: The case of Macao. *European Planning Studies, 22*(8), 1735–1743.

and geopolitics of the region. Despite Singapore being a small city state, its statesman Rajaratnam proclaimed decades ago that "the world is the hinterland of Singapore." Indeed, Singapore has earned a reputation as "a global city" over the past five decades.[62,63] Despite being a tropical country and located far away from the North Pole, Singapore has clearly articulated its interest in the Arctic, which is primarily economic and environmental in nature, given the low-lying topography of the island city state.

According to Bennett (2018),[64] Singapore officially maintains five concerns for the Arctic, which include "(1) Singapore's vulnerability to climate change as a low-lying island state and the Arctic's ability to serve as a bellwether; (2) the importance of upholding freedom of navigation across the world's oceans, including in the Arctic; (3) the potential for increased development in the Arctic to present new economic opportunities to Singapore; (4) the country's ability to contribute to a pool of knowledge about the Arctic; and (5) the fact that Singapore serves as a stopping point during the annual migration of Arctic shorebirds." Bennett also highlighted that the reasons for Singapore's interest in "that distant world of ice and snow" can be explained by the country's limited regional market and its government's stakes in shipping, maritime infrastructure, and global governance. Furthermore, given that Singapore's self-identity as a global city is closely linked to its strategic position in global transportation, it is important that the "city-state's polar pursuits also reflect the government's strategy of crafting a global national identity that is in step with its expansion of its overseas economic activities."[65]

Although the NSR has demonstrated its potential to change the landscape of the international economy, Singapore's leaders believe that the Route will only enhance trade in natural resources extracted from the Arctic region and is unlikely to profit the trade of commercial goods. The city state remains confident in its dominance in the global shipping

[62] Gu, X. H., Sheng, L., & Lei, C. K. (in press). *Specialization or diversification: A theoretical analysis for tourist cities*. Cities.

[63] Sheng, L., Gu, X. H., & Guo, H. Z. (in press). Business cycles of casino cities: Theoretical model, empirical evidence and policy implications. *Journal of Urban Affairs*.

[64] Bennett, M. M. (2018). Singapore: The "global city" in a globalizing Arctic. *Journal of Borderlands Studies, 33*(2), 289–310.

[65] Ibid.

industry and has little concern over the potential of the NSR to undermine the significance of the Straits of Malacca as well as its unique position as a global transport hub. At the same time, the development of the Arctic could provide new market opportunities for Singapore in sectors that are critical to its economy, such as its provision of offshore platforms and support vessels for Arctic conditions.[66] Bennett's[67] analysis supports this view: "as climate change accelerates, the Singaporean government's Arctic efforts suggest that it sees the increasingly maritime region as a new scalar fix for overseas investment that it is securing through unconventional partnerships while living up to its quest to view the world as its hinterland. Singapore's involvement in the Arctic may globalize the region's economy, but it may also deepen northern dependence on place-based sectors like natural resources and shipping."

Since the publication of the Nuuk criteria for observer states in the Arctic Council, the Singapore government has taken a keen interest in Arctic affairs and sought a position as an observer state in the Council. Singapore has also claimed to be a stakeholder in the Arctic on the basis of its national interests in climate change, maritime issues, and global governance. At the same time, Singapore has established comprehensive cooperative efforts with multiple Arctic states in recent years.

As noted by Storey, "on 15 May 2013, Singapore, together with its Asian neighbors China, Japan, South Korea, and India, was granted observer status to the Arctic Council. Since then, Singaporean officials have attended nearly every meeting, actively participated in several of the working groups and task forces, and delivered speeches at important annual Arctic conferences that bring together scientists, policy makers, security practitioners, businesspeople and academics. Quietly and modestly, Singapore is building its Arctic credentials, and people are noticing."[68]

Singapore has been proactive in making connections with the indigenous populations of the Arctic Circle via the provision of economic assistance and the establishment of lists of plant species for conversation

[66] Solli, P. E., Rowe, E. W., & Lindgren, W. Y. (2013). Coming into the cold: Asia's Arctic interests. *Polar Geography, 36*(4), 253–270.

[67] Bennett, M. M. (2018). Singapore: The "global city" in a globalizing Arctic. *Journal of Borderlands Studies, 33*(2), 289–310.

[68] Storey, I. (2016). Singapore and the Arctic: Tropical country, polar interests. In V. Sakhuja & K. Narula, *Asia and the Arctic* (pp. 63–74). Springer.

and forums for dialogue and conversation.[69] It has also rapidly initiated a series of research projects in the Arctic. These were reflected in the remarks by Singapore's Prime Minister in the 2016 Arctic Circle Greenland Forum held in Nuuk:

> In my view, our engagement in the Arctic revolves around what can be referred to as the science, technology, education and management framework, or STEM as I mentioned earlier... Singapore is still new to the Arctic and we are a small country with limited resources. But where possible, we have tried to work constructively with our friends and partners in the Arctic, and our universities are actually exploring the many possibilities of cooperating in the Arctic... the National University of Singapore will be signing a memorandum of understanding with the University of Alaska Fairbanks. Our establishment of a corporate lab between NUS and Keppel on Arctic studies is another example of our keenness to understand the Arctic and what is going on there. We are also interested to raise scientific and technological awareness of Arctic issues in our domestic context.

Prior to the Singapore government's more recent engagement in Arctic affairs, the country had already established economic ties with individual countries in the Arctic Circle (including Russia) in recent decades. Burke and Saramago suggested that Singapore has fostered its bilateral relations with Russia by investing in areas that are of importance to Russia, such as economic development and modernization. Although cultural and linguistic differences between Singapore and Russia have slowed the pace of progress in bilateral relations between the two countries, both remain open to pursuing a relationship. Moreover, in light of the sanctions that Russia faces from Western nations, Singapore has become an increasingly appealing partner.

Singapore's Arctic strategies, take, for example, the strategy for pursuing its economic interests in the Arctic, have proven to be relatively successful. In contrast to the "high-pitched tone" of China in the region, Singapore has cultivated a "benign image" in the Arctic Circle.[70] To be

---

[69] Ministry of Foreign Affairs Singapore. (2021, May 20). *Singapore's participation in the Arctic Council Ministerial Meeting.*

[70] Burke, D. C., & Saramago, A. (2018). With all eyes on China, Singapore makes its own Arctic moves. *The Conversation.*

more specific, this "micro-state" has successfully transformed its weakness in asserting itself internationally into a source of strength.[71] Burke and Saramago[72] insisted that the anticipated NSR is likely to redirect a substantial volume of maritime traffic away from the city state, undermining its position as an international shipping hub. This has motivated Singapore's involvement in the Arctic region, as Singapore aims to impede the development of the NSR and also reap more advantages from this Route to make up for anticipated losses.

7.3   Summary

The fundamental changes to the Arctic landscape in recent decades have attracted the attention of countries outside the Arctic Circle. The new players, which primarily include the European Union, China, Japan, South Korea, and Singapore, now actively participate in discussions on Arctic affairs. The European Union pays relatively little attention to the potential economic benefits that it can reap from the Arctic; its interest mainly relates to climate change, environmental problems, and its relationship with Arctic countries. This contrasts with the common interest of China, Japan, and South Korea in the region. Japan's main objective for intervening in Arctic affairs and cooperating with Arctic countries is to mitigate the environmental problem of rising sea levels, which threatens the country's long-term development and the safety of its people. The Arctic's marine resources also represent an important alternative food source for Japan. South Korea is drawn to the Arctic by the opportunities to lower shipping costs, diversify its energy sources, and advance its scientific research. Specifically, the country stands to benefit from the development of new shipping routes and the abundance of natural minerals in the Arctic, which will reduce its trading costs and increase its risk management capabilities in its energy imports. Although China has been a strong economic competitor of the above two Asian countries, the rise in sea levels has created opportunities for all three countries to cooperate on issues relating to the Arctic.

Overall, despite being geographically distant from the North Pole and harboring distinct national interests, China, Japan, South Korea, and

[71] Ibid.
[72] Ibid.

Singapore have a comparable interest in Arctic affairs stemming from similar concerns. On the one hand, China, Japan, and South Korea, and Singapore are all concerned about climate change and are motivated to cooperate in scientific research. Their scientists are likely to work together, share knowledge, and collectively explore the polar region to advance understanding of the environment. On the other hand, the potentially massive economic benefits brought on by developments in the Arctic may induce political conflicts. Each of these Asian countries desires to reap benefits and expand its international influence. The recent disputes between China and the United States may also trigger new conflicts in the Arctic. Japan and South Korea may need to take sides when Sino–American relations become intense. It is foreseeable that these two countries will become enemies of China, unless their relationships with the United States deteriorate substantially.

The dynamics of relationships between the four East Asian countries will be an important factor that will determine their strategies toward the Arctic. First, as observed in other studies of regional cooperation in East Asia, the complex historical contexts that connect these countries will directly decide their attitudes (i.e., whether they will choose to be cooperative or hostile). Second, given the increasingly fierce competition in the region, countries like China, Japan, and South Korea will attempt to gain economic dominance; this could strain their relations in Arctic affairs. Therefore, it is vital for all participants from East Asia to find common ground in their Arctic policies.

Aside from the obstacles that it faces from other countries, China is beset by various problems that need to be resolved before its interests can be realized in the Arctic. First, its extensive geographical distance from the region and lack of Arctic territory affords it little legitimacy in affairs and decisions pertaining to the Arctic. Thus, it must rely on Arctic countries to assert its position in Arctic affairs. Second, it will be disadvantaged in any military conflict with the United States in the Arctic, because of its lack of military facilities in the region. Third, Chinese enterprises have encountered numerous challenges in the unfamiliar deep-water and polar environment. Last, the value of Chinese investments in the Arctic remains uncertain; many researchers are concerned that these projects may have low profitability.

REFERENCES

Almazova-Ilyina, A. B., Vinogradov, A. D., Krasnozhenova, E. E., & Eidemiller, K. Y. (2020). National interests of Japan and its emerging Arctic policy. In *IOP Conference Series: Earth and Environmental Science*. IOP Publishing. https://iopscience.iop.org/article/10.1088/1755-1315/539/1/012048

Babin, J., & Lasserre, F. (2019). Asian states at the Arctic Council: Perceptions in Western states. *Polar Geography, 42*(3), 145–159.

Bennett, M. M. (2018). Singapore: The "global city" in a globalizing Arctic. *Journal of Borderlands Studies, 33*(2), 289–310.

Burke, D. C., & Saramago, A. (2018). With all eyes on China, Singapore makes its own Arctic moves. *The Conversation.* https://portal.findresearcher.sdu.dk/en/publications/with-all-eyes-on-china-singapore-makes-its-own-arctic-moves

Campbell, C. (2013). *China and the Arctic: objectives and obstacles.* US-China Economic and Security Review Commission staff research report. http://library.arcticportal.org/1677/1/China-and-the-Arctic_Apr2012.pdf

Chater, A. (2021). *What is Japan's Arctic interest? The Polar Connection report.* https://polarconnection.org/japan-arctic-interest/

Daemers, J. (2012). *The European Union in the Arctic: A Pole Position?* Bruges Regional Integration & Global Governance Papers. https://www.coleurope.eu/sites/default/files/research-paper/brigg_2012_daemers.pdf

EU Publication Office. (2018). *Strategy for the Baltic Sea region.* https://op.europa.eu/en/publication-detail/-/publication/11db6442-686b-11e8-ab9c-01aa75ed71a1/language-en

Gu, X., & Sheng, L. (2010). A sensible policy tool for Pareto improvement: Capital controls. *Journal of World Trade, 44*(3), 567–590.

Holroyd, C. (2020). East Asia (Japan, South Korea and China) and the Arctic. In *The Palgrave handbook of Arctic policy and politics* (pp. 319–332). Palgrave Macmillan.

Huebert, R. (2019). *Breaking the ice curtain? Russia, Canada, and Arctic security in a changing circumpolar world.* Canadian Global Affairs Institute report. https://www.researchgate.net/publication/334066768_Breaking_the_Ice_Curtain_Russia_Canada_and_Arctic_Security_in_a_Changing_Circumpolar_World

International Trade Administration (ITA). (2021). *China's energy data.* https://www.trade.gov/country-commercial-guides/china-energy

Jakobson, L. (2010). *China prepares for an ice-free Arctic.* SIPRI report. https://www.sipri.org/publications/2010/sipri-insights-peace-and-security/china-prepares-ice-free-arctic

Jin, D., Seo, W. S., & Lee, S. (2017). Arctic policy of the Republic of Korea. *Ocean & Coastal Law Journal, 22*(1), 85–96.

Lee, B. (2013). *Korea's Arctic policy: A Korean route towards the Arctic frontier.* http://library.arcticportal.org/1902/1/Arctic_Policy_of_the_Republic_of_Korea.pdf

Ministry of Foreign Affairs Singapore. (2021, May 20). *Singapore's participation in the Arctic Council Ministerial Meeting.* https://www.mfa.gov.sg/Newsroom/Press-Statements-Transcripts-and-Photos/2021/05/20210521-SG-Arctic-Council-Ministerial-Meeting

Ng, A. (2021). *Tensions likely to grow as China seeks a bigger role in the Arctic.* CNBC. https://www.cnbc.com/2021/05/20/tensions-likely-to-grow-as-china-seeks-a-bigger-role-in-the-arctic.html

Offerdal, K. (2011). The EU in the Arctic: In pursuit of legitimacy and influence. *International Journal, 66*(4), 861–877.

Ohnishi, F. (2016). Japan's Arctic policy development: From engagement to a strategy. In *Asian countries and the Arctic future* (pp. 171–182). Springer.

Østhagen, A. (In press). *The New Geopolitics of the Arctic: Russia, China, and the EU.* European Review.

Park, Y. K. (2014). South Korea's interests in the Arctic. *Asia Policy, 18*, 59–65.

Pelaudeix, C. (2018). Along the road: China in the Arctic. *EUISS 2018* (13). https://www.iss.europa.eu/sites/default/files/EUISSFiles/Brief%2013%20Arctic.pdf

Perez, E. C., & Yaneva, Z. V. (2016). The European Arctic policy in progress. *Polar Science, 10*(3), 441–449.

Pincus, R. (2020). Three-way power dynamics in the Arctic. *Strategic Studies Quarterly, 14*(1), 40–63.

Sakhuja, V., & Narula, K. (Eds.). (2016). *Asia and the Arctic: Narratives, perspectives and policies.* Springer.

Sheng, L. (2014a). Capital controls and international development: A theoretical reconsideration. *Global Policy, 5*(1), 114–120.

Sheng, L. (2014b). Economic structure, cost outsourcing and global imbalances. *Journal of Australian Political Economy, 74*, 81–94.

Sheng, L. (2014c). The effects of foreign expansion on local growth: The case of Macao. *European Planning Studies, 22*(8), 1735–1743.

Sheng, L. (2015). Theorizing income inequality in the face of financial globalization. *The Social Science Journal, 52*(3), 415–424.

Sheng, L. (2021). *How Covid-19 reshapes new world order: Political economy perspective.* Springer.

Sheng, L., & do Nascimento, D. F. (2021a). *The Belt and Road Initiative in South-South Cooperation: The impact on world trade and geopolitics.* Palgrave Macmillan.

Sheng, L., & do Nascimento, D. F. (2021b). *Love and trade war: China and the US in historical context.* Springer.

Solli, P. E., Rowe, E. W., & Lindgren, W. Y. (2013). Coming into the cold: Asia's Arctic interests. *Polar Geography, 36*(4), 253–270.

Stępień, A., & Koivurova, T. (2017). *Arctic Europe: bringing together the EU Arctic policy and Nordic cooperation.* Publications of the Government's analysis, assessment and research activities 15/2017. https://julkaisut.valtioneu vosto.fi/handle/10024/160217

Storey, I. (2016). Singapore and the Arctic: Tropical country, polar interests. In V. Sakhuja & K. Narula, *Asia and the Arctic* (pp. 63–74). Springer.

The Headquarters for Ocean Policy. (2015). *Japan's Arctic Policy.* https://www8.cao.go.jp/ocean/english/arctic/pdf/japans_ap_e.pdf

Tonami, A. (2014). The Arctic policy of China and Japan: Multi-layered economic and strategic motivations. *The Polar Journal, 4*(1), 105–126.

Tonami, A. (2021). *UK–Japan responses in the Arctic.* Chatham House report. https://www.chathamhouse.org/2021/03/security-frontier/uk-japan-res ponses-arctic

Wang, X. Q. (2020). An analysis of the law of China-US-Russia Triangle Relations. *Studies in Russia, Eastern Europe and Central Asia, 2020*(3), 32–52.

Xia, H. (2019). *China's polar icebreaker Xuelong 2 finished its icebreaking tasks.* XinhuaNet. http://www.xinhuanet.com/english/2019-11/25/c_1385 81294.htm

Yao, Z. (2019). Ice Silk Road framework welcomed by countries, sets new direction for Arctic cooperation. *Global Times.* https://www.globaltimes.cn/con tent/1144928.shtml

Yin, Y., & Sheng, L. (2021). Theorizing about global imbalances: An inequality perspective. *Argumenta Oeconomica, 46*, 169–181.

Yoon, S. (2016). A cooperative maritime capacity-sharing strategy for the Arctic region: The South Korean perspective. In V. Sakhuja & E. Narula, *Asia and the Arctic* (pp. 49–62). Springer.

Zhang, D. M., Zhao, S. Y., & Zhang, D. Z. (2020). An analysis of the "New Northern Policy" of South Korea's Moon Jae-in Government. *Journal of Liaoning University* (Philosophy and Social Sciences Edition).

Zhao, J. (2013, January 21). *Speech by Ambassador Zhao Jun at the Arctic Frontiers Conference.* https://www.rcinet.ca/eye-on-the-arctic/2013/02/05/arc tic-frontiers-russian-voices/

The Polar Silk Road and the Belt and Road Initiative: Integration and Optimization

Abstract The Belt and Road Initiative presents a realistic demand for expansion and connectivity to the entire Arctic region. As a significant component of the Belt and Road Initiative, China's proposed Polar Ice Silk Road sets out a vision for China's future maritime cooperation and geopolitical strategy in the Arctic. This chapter addresses concerns about China's Polar Silk Road and Belt and Road Initiative and their practical effects on the Arctic region.

Keywords Multilateral cooperation · Belt and Road Initiative · Polar Silk Road

Where China's Arctic policies and its cooperation with Russia in the North Pole are concerned, several concepts such as the Polar Silk Road and the BRI can be confusing and difficult to interpret. In this chapter, we distinguish these concepts and clarify their content. To date, China's Polar Silk Road has been the country's only initiative dedicated to maritime cooperation and geopolitical competition in the Arctic. However, the content of this project remains unclear and it lacks specific policies to guide its implementation. This chapter shows that the BRI creates realistic demands for the expansion of transport networks to connect the Arctic region, which in turn will facilitate China's processes of "docking" and "landing" with the international community—that is, its cooperation with other countries and the implementation of its Arctic initiative, respectively. In addition, because China's geoeconomic cooperation with other countries has been continually integrated into its planning, the Arctic countries' ambiguity regarding this concept has hampered efforts by Arctic countries and the international community to understand China's Arctic policy. This has led to misunderstandings and misinterpretations. To provide clarity on this topic, this chapter restores the history of the proposed Polar Silk Road initiative, clarifies its concepts, and explores the vision of Sino–Russian cooperation on the Polar Silk Road and the challenges it faces.

8.1   Features of Continuity and Complementarity Between the Polar Silk Road and the BRI

The BRI, China's grandest cooperation initiative in recent decades, expresses the country's ambition and determination to uphold its basic state policy of opening up. The BRI emerged from the combination of two previous policies—the Silk Road Economic Belt and the 21st Century Maritime Silk Road. The former was proposed in 2013 during President Xi's visit to Kazakhstan and the latter was proposed in Indonesia in 2013 during a speech in Indonesian Parliament. In the same year, China joined the Arctic Council as an observer state. In March 2015, the Chinese government released a document titled Vision and Actions on Jointly Building the Silk Road Economic Belt and 21st Century Maritime Silk Road, which aimed to synchronize its development plans and promote joint actions among countries along the 21st Century Maritime Silk Road.

In June 2017, China's NDRC and SOA jointly released another document, Vision for Maritime Cooperation under the Belt and Road Initiative.[1] It was in this particular document that the Polar Silk Road was first proposed. In essence, the proposed Polar Silk Road fell within the framework of the BRI. However, in the full text of the document, the term "Polar Silk Road" was not mentioned. Instead, the document called for the joint construction of a "Blue Economic Corridor connecting

---

[1] National Development and Reform Commission and the State Oceanic Administration of the People's Republic of China. (2017). Vision for maritime cooperation under the Belt and Road Initiative.

Europe via the Arctic Ocean" as part of the BRI. China's Arctic initiative of Blue Economic Corridor called for multilateral governance in the Arctic, international scientific investigation, regional environmental protection, the exploration of maritime routes in the Arctic Ocean, the sustainable development of Arctic resources, and other issues. It was the first time that China proposed the development of a "blue economic passage" and maritime trade to link Europe across the Arctic Ocean, and served as the predecessor of the Polar Silk Road. In the same year, when China's President Xi met with visiting Russian Prime Minister Dmitry Medvedev, the two countries formally proposed the joint construction of the Polar Silk Road. This was the first time that Chinese leaders expressed to the international community that China was motivated to carry out strategic docking with the relevant countries and to jointly build the Polar Silk Road in the Arctic.

In January 2018, China's Ministry of Foreign Affairs released a white paper titled "China's Arctic Policy." This was the first document in which China provided full details of its identity, responsibilities, principles, and policy positions in affairs concerning the Arctic:

China is an important stakeholder in Arctic affairs. Geographically, China is a "near-Arctic state," one of the continental states that are closest to the Arctic Circle. The natural conditions of the Arctic and their changes have a direct impact on China's climate system and ecological environment, and, in turn, on its economic interests in agriculture, forestry, fisheries, the marine industry and other sectors.

China stands for steadily advancing international cooperation in the Arctic. It has worked to strengthen such cooperation under the Belt and Road Initiative according to the principles of extensive consultation, joint contribution and shared benefits. It has emphasized policy coordination, infrastructure connectivity, unimpeded trade, financial integration and closer people-to-people ties. Concrete steps for cooperation include coordinating development strategies with the Arctic States, encouraging joint efforts to build a blue economic passage linking China and Europe via the Arctic Ocean, enhancing digital connectivity in the Arctic, and building a global infrastructure network. China hopes to work for the common good of all parties and further common interests in the Arctic.

The Silk Road Economic Belt and the 21st-century Maritime Silk Road (Belt and Road Initiative), important cooperation initiatives of China, will bring opportunities for parties concerned to jointly build a "Polar Silk Road," and facilitate connectivity and sustainable economic and social development in the Arctic.[2]

According to the timing of the Polar Silk Road's implementation and its goals, this initiative was actually designed as a natural extension of the BRI to the North Pole. In other words, it was both complementary to the BRI and designed to optimize the BRI. Indeed, The Chinese government has identified the Polar Silk Road as a significant component of the BRI.[3] According to Yang, the Polar Silk Road is a project developed by China that aims to promote international cooperation between specific Arctic countries.[4] In this regard, it fits within the overall framework of the BRI and builds on China's previous experience in international cooperation to address its economic and political needs.

As reflected in the 2018 white paper, China's ambitions in constructing the Polar Silk Road are based on its broader strategic vision and active foreign strategy. In particular, its new foreign policy approach toward the Arctic reveals its ambitions on the international stage. By calling for multilateral cooperation in the North Pole, China wishes to expand its influence beyond that of its immediate territory and the regions in its periphery (e.g., Southeast Asia).[5] Given the increasing uncertainty in the international political situation owing to its intensifying rivalry with the United States, China deems it vital to seek as many alternatives as possible to ensure its core interest—the secure trade and supply of energy.[6,7] In

[2] State Council Information Office of the People's Republic of China. (2018). China's Arctic Policy.

[3] Jiang, Y. A. (2019). Multilateral cooperation on the Polar Silk Road: Opportunities, challenges and development paths. *Pacific Journal* (Taipingyang Xuebao), *27*(8), 67–77.

[4] Yang, J. (2018). The international environment and response to the construction of the Polar Silk Road. *People's Forum· Academic Frontier* (Renmin Luntan·Xueshu Qianyan) (11), 13–23.

[5] Sheng, L. (2021) *How Covid-19 reshapes new world order: Political economy perspective.* Springer.

[6] Sheng, L. (2014). Economic structure, cost outsourcing, and global imbalances. *Journal of Australian Political Economy, 74*, 81–95.

[7] Sheng, L. (2015). Theorizing income inequality in the face of financial globalization. *The Social Science Journal, 52*(3), 415–424.

this context, China is determined to take advantage of the Polar Silk Road to gain access to the Arctic region's rich energy resources.[8] For this reason, it is unlikely that China will exclude itself from Arctic affairs. Instead, by identifying its position as a "near Arctic-state" and appealing for good governance in the region, China is attempting to establish the legitimacy of its right and obligation in affairs concerning this region.[9]

According to China's Arctic White Paper 2018, the construction of the Polar Silk Road mainly involves two components: the development of resources in the North Pole and the utilization of trade routes. More specifically, these refer to the establishment of three maritime routes in the Arctic Ocean[10]:

(1) The Northern Sea Route (NSR): [...] the only route currently used for long-distance commercial shipping and a sub-section of the Russian part of the "North East Passage," stretching from Norway to the Bering Strait. It includes sea waters (comprising internal sea waters, territorial sea, a contiguous zone and exclusive economic zone) from the Novaya Zemlya archipelago in the West, to the Bering Strait in the East.

(2) The Northwest Passage (NWP): the NSR and the sections are regarded by Russia and Canada as internal waters, and not as international straits, meaning that no ship can navigate through them without their consent.

(3) The Central Passage, through the central sea area of the Arctic Ocean.

The three routes provide shortcuts that largely reduce the travelling distance between North America, East Asia, and Western Europe. Of the three routes, the NSR has the greatest geopolitical significance. Ships using the NSR can take advantage of the port clusters, navigation systems, icebreaker fleets, and infrastructure related to maritime transport that

---

[8] Li, Z. F., & Peng, Y. (2018) "The Polar Silk Road" and the Great Arctic Network: Role, evolution and China's strategy. *Northeast Asian Economic Research, 2*(5), 5–20.

[9] Lim, K. S. (2018). China's Arctic policy and the Polar Silk Road Vision. *Arctic Yearbook,* 2018, 420–432.

[10] Pelaudeix, C. (2018). Along the Road China in the Arctic. Connectivity and security. Brief, 13.

have been developed along this route. In addition, given the significant global warming that has taken place leading to the NSR becoming advantageous. The NSR occupies a key position in global shipping; it connects China's seas and the Eurasian region to the Pacific Ocean and the western border of Russia.[11] The NSR is also important in the context of increasingly uncertain Sino–American relations. All other sea routes are controlled by China's main geopolitical rival, the United States, and therefore cannot be regarded as permanently stable, secure, and reliable routes in the long term. In contrast, the NSR provides a secure maritime channel for Sino–Russian cooperation and is the only safe passage recognized by the Shanghai Cooperation Organization. The NSR can also be incorporated into China's Silk Road Economic Belt owing to its high operational reliability, great potential transport capacity, and overall attractiveness. A comprehensive assessment of the NSR and other transport corridors was undertaken during a Russian–Chinese research project that was conducted by the Russian Academy of Military Sciences and the Russian–Chinese Center for Regional Cooperation Studies at Jilin University. It was eventually found that the NSR represented the most important element of a multiform intermodal transport system, with great potential for sustainable development and improved security of related parties. The seminar recommended both countries to explore and consider other features of the NSR, in addition to their seasonal dependence on the route.[12]

Expanding the BRI and connecting states within the Arctic region is a key objective of China. Biedermann[13] noted that China's primary goal in building the Polar Silk Road ultimately depends on the factors and conditions under which it seeks to influence and manage its Arctic affairs. Furthermore, as one of the world's largest producers, consumers, and importers of energy, China has long been exposed to threats to its energy security, such as during the Malacca dilemma as a control of the Strait

[11] Blunden, M. (2012). Geopolitics and the northern sea route. *International Affairs,* *88*(1), 115–129.

[12] Zhu, X. P., Zhang, Y. F, Liu, X., & Wang, Y. G. (2018) Implementing the spirit of the 19th National Congress to build the "Ice Silk Road"-Jilin University-Russian Academy of Military Sciences "Ice Silk Road" Seminar. *Northeast Asia Forum, 27*(2), 3–7.

[13] Biedermann, R. (2020). The Polar Silk Road: China's multilevel Arctic strategy to globalize the Far North. *Contemporary Chinese Political Economy and Strategic Relations,* *6*(2), 571–615.

of Malacca by anyone will also mean that they control the oil routes to China and thus the economy too indirectly. Increasing Sino-US trade friction also offers Russia an opportunity to enhance cooperation with China.[14] Recently, China and Russia have positively decided to jointly develop the Polar Silk Road, especially the NSR. The two countries have also expanded their economic cooperation in the Arctic, such as through the Yamal LNG project, the world's largest liquefied natural gas project, which helps to meet China's growing energy needs and Russia's interest in Arctic resources.[15] Moreover, the establishment of Arctic shipping routes such as the NSR will allow ships travelling to and from China to avoid the lengthy and potentially more volatile routes from the country's southwest across the South China Sea and the Indian Ocean. The NSR also offers additional advantages, such as direct transport links to major European ports and shorter delivery times.[16] China has always been motivated to develop new shipping routes from the Arctic to Europe to establish a closer relationship with the EU.[17]

8.2 The Polar Silk Road and China's Multilateral Cooperation in Arctic Affairs

As an outsider to the North Pole, China is considered a "geographically disadvantageous country" in its affairs.[18] This leaves China little choice but to get involved in Arctic affairs by enhancing its cooperation with Arctic states. As a non-Arctic country, China's proposed Polar Silk Road follows the basic principles of the BRI, that is, seek cooperation with other countries through multilateral cooperation, construction, and sharing.

---

[14] Sheng, L., & do Nascimento, D. F. (2021). *Love and trade war: China and the US in historical context*. Springer.

[15] Ministry of Economic Development of the Russian Federation. (2021). Russia boosts Arctic LNG shipments to China to meet growing energy demand. Russia boosts Arctic LNG shipments to China to meet growing energy demand.

[16] Yilmaz, S. (2017). Exploring china's arctic strategy: Opportunities and challenges. *China Quarterly of International Strategic Studies, 3*(1), 57–78.

[17] Blaxekjær, L. Ø., Lanteigne, M., & Shi, M. (2018). The Polar Silk Road & the West Nordic Region. *Arctic Yearbook*, 2018, 437–455.

[18] Li, X. P., Fang, Z., & Liu, H. Y. (2017). Rights and obligations of the Arctic Region under the United Nations Convention on the Law of the Sea and their relationship with geographically disadvantaged States. *Polar Research* (2), 279–285.

Nonetheless, when China attempted to establish its own Arctic identity, these actions raised suspicion among the Arctic region's more territorial and possessive states.[19] Woon claimed that although the Polar Silk Road (and more broadly the BRI) projects currently generate substantial geopolitical skepticism and resistance from a wide range of local actors, most Chinese scholars hold a positive view.[20] Scholars have argued that the inclusiveness and mutual benefits that the BRI has demonstrated previously will help other parties accept China's actions in Arctic affairs. Furthermore, they have suggested that over time, China's sincerity and the economic benefits of its initiatives will facilitate the forging of partnerships in the Arctic. To this end, China has invested extensive resources to demonstrate its sincerity and form partnerships with Arctic countries.

Of the eight Arctic countries, China has worked most closely with Norway, Finland, Demark, and Iceland. In 2012, China and Iceland signed the Framework Agreement between China and Iceland on Arctic Cooperation.[21] In December 2013, China and the Nordic countries (Iceland, Denmark, Finland, Norway, and Sweden) signed the Cooperation Agreement on the China–Nordic Arctic Research Center (CNARC).[22] In 2018, Chinese investors were invited to participate in the construction of Finland's Arctic Corridor tunnel.[23] China has also worked with Norway and Finland in the construction of infrastructure in the Arctic. For instance, the three countries have partnered to build the Finnish–Norwegian Arctic Railway, which will run between the NSR and Europe. Increasingly, China has invested in Iceland's development of technologies for the extraction of natural resources.[24]

---

[19] Bennett, M. M. (2015). How China sees the arctic: Reading between extraregional and interregional narratives. *Geopolitics, 20*(3), 645–668.

[20] Woon, C. Y. (2020). Framing the "Polar Silk Road" (bingshang sichou zhilu): Critical geopolitics, Chinese scholars and the (re)positionings of China's arctic interests. *Political Geography, 78*, 102141.

[21] Embassy of the People's Republic of China in the Republic of Iceland. (2012, April 25). Chinese premier Wen Jiabao pays official visit to Iceland.

[22] Arctic Portal. (2013, December 11). *China–Nordic Arctic Research Center*. Retrieved September 14, 2021, from https://arcticportal.org/ap-library/news/1136-china-nordic-arctic-research-center-inaugurated.

[23] Chinadaily. (2018, March 2). Finland touts "Arctic Corridor" tunnel.

[24] Reinke de Buitrago, S. (2020). China's aspirations as a "Near Arctic State": Growing stakeholder or growing risk? In *Handbook on geopolitics and security in the Arctic* (pp. 97–112). Springer.

Finally, China has channeled substantial investments to Greenland, including investments in its economy, infrastructure, land, mineral extraction, and tourism sector.[25] Havnes noted that although the eight Arctic countries have different perspectives from China on the subject of Arctic governance, most of these states have displayed positive attitudes in their engagements with China on Arctic affairs.[26] Furthermore, Zheng noted that the eight Arctic countries, especially the Nordic countries, which are the main players in Arctic affairs, have more accommodating attitudes toward China.[27] This contrasts with the attitudes of the United States and Canada, which generally resist any participation from external parties in Arctic affairs.

Greenland is the world's largest and northernmost land mass. As it is located at the intersection of the Arctic Ocean and the North Atlantic Ocean, where it connects the Northeast, Northwest, and Central waterways, Greenland occupies a very important strategic position in the Arctic region. For this reason, effective cooperation between China and Greenland is paramount to advancing the construction of the Polar Silk Road.

China is incentivized to build (or in some cases develop) several airports in Greenland, including airports in Nuuk, Ilulissat, and Qaqortoq.[28] Greenland's rich natural resources have also attracted the interest of China. For instance, Chinese government and industry partners have invited Greenland to participate in the annual China Mining Conference.[29] China has also been involved in various mining projects in Greenland. Four mining projects—at Kvanefield, Citronen Fjord, Isua, and Wegner Halvo—have garnered much interest from China as well as multiple political observers. For instance, although the Kvanefield project is wholly controlled by Greenland Minerals Ltd. (GML), one of China's

[25] Ibid.

[26] Havnes, H. (2020). The Polar Silk Road and China's role in Arctic governance. *Journal of Infrastructure, Policy and Development*, 4(1), 121–138.

[27] Zheng, Y. Q. (2019). China-Nordic blue economic passage: Basis, challenges and paths. *China International Studies* (5), 29–49.

[28] Reinke de Buitrago, S. (2020). China's aspirations as a "Near Arctic State": Growing stakeholder or growing risk? In *Handbook on geopolitics and security in the Arctic* (pp. 97–112). Springer.

[29] Mohr, J. (2020). China in the Arctic and the case of Greenland. In *Handbook on geopolitics and security in the Arctic* (pp. 113–129). Springer.

largest state-owned enterprises, Shenghe Resources Holding, has invested in GML and holds 11% of its shares.[30] In addition, GML has formed a partnership with Shenghe Resources Holding to undertake environmental surveys and investigations for mining projects.[31]

China's increasing involvement in Greenland has complicated its relations with Denmark, a country with territorial claim over Greenland and which is especially motivated to maintain its control over the region. It is precisely Greenland's position in the Arctic that affords Denmark its status as an Arctic country and provides Denmark with some influence on Arctic affairs. However, Greenland seeks to gain greater economic independence through its cooperation with China in the Polar Silk Road initiative, to obtain sufficient capital for negotiating its independence from the Danish government. Although Denmark maintains an open attitude toward most Chinese investment under the BRI, it is very sensitive and resistant to China's involvement in Greenland affairs.

China also recognizes the importance of developing dialogue and cooperation with non-Arctic countries on polar affairs. It has, for instance, developed such a dialogue with the United Kingdom and France. In 2016, China, Japan, and South Korea launched a high-level dialogue to promote communication and exchange between the three countries and strengthen their cooperation in scientific research and commercial sectors pertaining to the Arctic. To increase its political influence, status, and right to speak on Arctic affairs, China is likely to continue to work within the current system of Arctic governance and achieve its objectives by actively pursuing scientific and economic cooperation with its partners in the region.[32]

Nonetheless, the European Union has expressed multiple security concerns over China's increasing presence in the Arctic.[33] First, there is a risk that China's Polar Silk Road will amplify the fragmentation of the European Union's political cohesion and undermine its strategic autonomy in the region. Second, member states of the European Union are deeply concerned that China will gain control of transport and

---

[30] Ibid.

[31] Ibid.

[32] Kossa, M. (2019). China's Arctic engagement: Domestic actors and foreign policy. *Global Change, Peace and Security, 32*(1), 19–38.

[33] Pelaudeix, C. (2018). Along the road: China in the Arctic. Connectivity and security. Brief, 13.

logistics hubs in the Arctic, and potentially even militarize the region. Third, the rapid and large-scale construction of the Polar Silk Road may undermine the European Union's environmental and social standards. Concerned about China's multilateral cooperation in Arctic affairs, Borgerson[34] argued that the United States should take the lead in developing diplomatic solutions that can address the competing needs and potential conflicts between Arctic nations. Otherwise, the Arctic could erupt into a "frenzied armed scramble for resources."

Several scholars have viewed China's strategic cooperation with Russia as its most effective thus far; indeed, it has been used as a general framework for various BRI-related policies involving countries in Europe and Asia. Although China is seeking more in-depth bilateral and multilateral opportunities for cooperation with other Eurasian countries and institutions on its BRI, these arrangements have had limited success.[35]

Although China and Russia have had a history of successful cooperation, they differ in their views concerning the development of the Arctic. Russia prefers to cooperate bilaterally with China, to better safeguard its rights as an Arctic state. However, China has made clear its intention to work with all parties in the construction of the Polar Silk Road. The diverging interests of the two countries may also create some uncertainty and barriers to cooperation. For instance, although Russia welcomes China's huge investments in the region, it avoids sharing its advanced technologies with China owing to its concerns over China's engagement in Arctic affairs.[36] Sino–Russian cooperation may essentially be characterized as "Trust and Control."[37] However, Weber argued that Russia is more dependent on China owing to the latter's abundance of financial

[34] Borgerson, S. (2008). Arctic Meltdown: The economic and security implications of global warming. *Foreign Affairs, 87*(2), 63–77.

[35] Yilmaz, S., & Liu, C. M. (2019). Remaking Eurasia: The Belt and Road Initiative and China-Russia strategic partnership. *Asia Europe Journal, 18*(3), 259–280.

[36] Alexeeva, O., & Lasserre, F. (2018). An analysis on Sino-Russian cooperation in the Arctic in the BRI era. *Advances in Polar Sciences, 29*(4), 269–282.

[37] Pincus, R. (2020). Three-way power dynamics in the Arctic. *Strategic Studies Quarterly, 14*(1), 40–63.

and economic resources.[38],[39] Since 2013, China and Russia have been engaged in dialogue on Arctic affairs and cooperated in the construction of the Polar Silk Road. In December 2017, the Yamal LNG project was launched, cementing an important milestone in Sino–Russian cooperation on the Polar Silk Road.[40] Hsiung[41] argued that the significance of the Yamal LNG project may be overstated, as the project is unlikely to turn a profit unless it receives full support from the Russian government. Although the profitability of the Yamal LPG project remains to be determined, it represents a significant step for Sino–Russian relations and China's participation in Arctic affairs.

8.3   THE INTENSIFICATION OF CHINA'S COOPERATION IN NON-TRADITIONAL SECURITY IN THE ARCTIC

Issues that relate to traditional security, such as military security, generally receive substantial public interest. However, non-traditional security—such as human security, social security, environmental security, and cultural security[42]—has generally been overlooked. Issues relating to non-traditional security strongly influence the lives of Arctic people. For instance, research has suggested that the melting of ice brought about by pollution will make food resources in the Arctic less accessible to its local populations, threatening their survival. Additionally, changes to and parties responsible for the Arctic's economic structure have resulted in its labor market failing to provide sufficient jobs for younger people, hampering economic and social development in the region. As many young people in the Arctic leave their countries in search of opportunities elsewhere, economic and social problems have worsened. Such threats to

[38] Weber, J. (2020). Limited cooperation or upcoming alliance? Russia, China and the Arctic. In *Handbook on geopolitics and security in the Arctic* (pp. 345–361). Springer.

[39] Sheng, L., & Share, M. (2022). Exchange between Russia and the Guangdong–Hong Kong–Macao Greater Bay Area in a historical context. *Revista de Cultura, 67*, 46–59.

[40] Yearender: Growing China brings prosperity to Eurasia. *Xinhua* (2017, December 31).

[41] Weidacher Hsiung, C. (2016). China and Arctic energy: Drivers and limitations. *The Polar Journal, 6*(2), 243–258.

[42] Havnes, H. (2020). The Polar Silk Road and China's role in Arctic governance. *Journal of Infrastructure, Policy and Development, 4*(1), 121–138.

the Arctic's non-traditional security—in this case, its economic and social security—continuously create problems in the region.

China's extensive participation in Arctic affairs has been followed by its gradually increasing cooperation in non-traditional security. As mentioned, research and state policies concerning Arctic security have tended to focus on issues of traditional security and often neglected issues of non-traditional security, such as various socio-environmental and socio-economic factors.[43] Nonetheless, risks to the Arctic's non-traditional security—which may, for instance, come in the form of oil spills and rescue operations—are likely to increase with the accelerating commercial development in the region.[44] Ho suggested that greater investments in infrastructure and the provision of maritime services (while minimizing environmental impacts) will be key to ensuring safe and reliable transportation in the region before the NRS can be reliably used as a shipping route between Europe and Asia.[45]

The issue of environmental security has been a key concern of the Arctic Council since its founding in 2012. Rather than "ecologism," as a new political ideology based on the position that the non-human world is worthy of moral consideration the Arctic's environmental issues can be conceptualized as "environmentalism," which emphasizes the relationship between human beings and nature.[46] Discovering the means for human beings to coexist harmoniously with nature is a key objective of sustainable human development. The Arctic states have mutually declared that the Arctic environment should be preserved and that sustainable development should be a top priority.[47] This represents one of the fundamental

---

[43] Hossain, K., Zojer, G., Greaves, W., Roncero, J., & Sheehan, M. (2017). Constructing Arctic security: An inter-disciplinary approach to understanding security in the Barents region. *Polar Record*, *53*(1), 52–66.

[44] Havnes, H. (2020). The Polar Silk Road and China's role in Arctic governance. *Journal of Infrastructure, Policy and Development*, *4*(1), 121–138.

[45] Ho, J. (2010). The implications of Arctic sea ice decline on shipping. *Marine Policy*, *34*(3), 713–715.

[46] Dobson, A. (2007). *Green political thought*. Routledge.

[47] Arctic Council. (1996). Declaration on establishment of the Arctic Council (The Ottawa Declaration) 1996.

principles on which the Arctic Council was originally established, as environmental security is vital to the prosperity of human beings.[48] All foreign countries that seek to participate in Arctic affairs must follow the commitments outlined in the Declaration on the Establishment of the Arctic Council. This document states that any foreign state wishing to participate in Arctic affairs must have the capacity to help Arctic countries to address threats to their non-traditional security.

Icebreaking has created numerous economic opportunities benefiting relevant parties in the Arctic. However, it has also put countries positioned in low-lying coastal areas at risk. One unintended consequence of icebreaking is the rising coastal sea levels, which cause flooding in many countries. Such events threaten human security, which is a form of non-traditional security. Research has shown that the melting of sea ice in the Arctic can lead to haze and air pollution in eastern China, which will endanger the health of the local population.[49,50] Climate change not only represents a threat to Arctic countries and other countries in low-lying coastal areas but also threatens human, social, and economic security worldwide. China's Arctic policy indicates that "the developing situation in the Arctic now extends beyond the original inter-Arctic states. It relates to the interests of countries outside the region, the interests of the international community as a whole, as well as the survival, development, and shared future of mankind."[51] In this regard, Arctic issues are not confined to the region but concern the whole world. Furthermore, China's interest in environmental security is also political. Global warming can potentially undermine China's social stability and erode the legitimacy of the ruling Chinese Communist Party.[52] These questions have already seemed more

[48] Greaves, W. (2016, March 22). *Thinking critically about security and the Arctic in the Anthropocene.* The Arctic Institution.

[49] Zou, Y., Wang, Y., Zhang, Y., & Koo, J. H. (2017). Arctic sea ice, Eurasia snow, and extreme winter haze in China. *Science Advances, 3*(3), e1602751.

[50] Wang, H. J., Chen, H. P., & Liu, J. P. (2015). Arctic sea ice decline intensified haze pollution in eastern China. *Atmospheric and Oceanic Science Letters, 8*(1), 1–9.

[51] State Council Information Office of the People's Republic of China. (2018). China's Arctic Policy.

[52] Kopra, S. (2020). China, great power responsibility and Arctic security. In *Climate change and Arctic security* (pp. 33–52). Springer.

and more important with surging social inequality in China.[53] Therefore, in the interest of promoting non-traditional security, China is motivated to cooperate with Arctic countries and actively participates in the Arctic Council.

China actively engages in non-traditional security cooperation in the Arctic to build a new, greener BRI in the region.[54] On April 26, 2017, the Chinese Ministry of Environmental Protection (MEP) proposed new guidelines for the construction of a new and sustainable BRI. Specifically, it expressed China's desire to cooperate with Arctic countries on issues pertaining to eco-environmental protection and risk management, support collaboration between governments and enterprises, promote sustainable trade and green infrastructure, and develop a green financial system.[55] Overall, these approaches demonstrate China's willingness to cooperate with Arctic countries in strengthening the region's capacity to deal with environmental risks. In terms of practical efforts, China has participated in several environmental research projects, such as the International Polar Year Program, The Ny-Alesund Science Managers Committee, and the International Arctic Science Committee.[56] In addition, China has dedicated substantial resources to Arctic exploration. For instance, it developed its icebreakers, "Xue Long" and "Xue Long 2," for scientific research and environmental exploration.[57] Xue Long 2 completed several missions in the Arctic, involving the supply of materials, ecological investigations, and scientific expedition.[58] This icebreaker is not only used for scientific research and commercial purposes but also for rescue operations and to ensure the safety of Arctic researchers. Because of the extreme environmental conditions in the Arctic, ice must first be

[53] Yin, Y. C., & Sheng, L. (2021). Theorizing about global imbalances: An inequality perspective. *Argumenta Oeconomica, 46.*

[54] Liu, N. (2018). Will China build a green Belt and Road in the Arctic? *Review of European, Comparative & International Environmental Law, 27*(1), 55–62.

[55] Ministry of Ecology and Environment of the People's Republic of China. (2017). Guidance on promoting Green Belt and Road. Retrieved September 26, 2021, from.

[56] Campbell, C. (2013). China and the Arctic: Objectives and obstacles. US-China. Economic and Security Review Commission.

[57] XinHuaNet. (2019, November 25). China's polar icebreaker Xuelong 2 finished its icebreaking tasks.

[58] XinHuaNet. (2021, May 7). Chinese research icebreaker Xuelong 2 completes Antarctic expedition.

crushed before infrastructure or scientific research centers can be reached. As a result, icebreakers are important in facilitating scientific progress in the Arctic. Moreover, icebreakers can help free ships stuck in the ice. Finally, icebreakers are useful for transferring supplies from other regions when access to normal cargo is limited, especially in an emergency.

Arctic countries are also concerned about human security. The extraction of fossil fuels has created opportunities and problems for the community security of Arctic people, their identities, their local cultures, the regional economy, and their general living and health conditions.[59] The abundance of fossil fuels in the Arctic has caused local communities to change their economic structure, as support for the agricultural, fishing, and hunting sectors has decreased.[60] On the one hand, natural resources, such as oil and gas, provide economic benefits. On the other hand, the extraction and use of these resources undermine domestic living standards; it alters the Arctic's social and economic structure as well as its plans for long-term development. Therefore, sustainable development, which includes human security and social security, is the primary goal of Arctic countries. Hossain, Martín, and Petretei[61] noted that improvements in human security can simultaneously advance social security, including security in the areas of culture, religion, identity, and traditional customs.

To form sustainable, mutual, and long-term partnerships with Arctic countries, China will need to intensify its cooperation in issues relating to human and social security. In its Arctic policy white paper, the Chinese government acknowledges the importance of sustainable development and expresses its desire to construct a peaceful and stable region with stakeholders in the Arctic, improve the living standards of local inhabitants, and protect the Arctic's unique ecosystem and biodiversity.[62] From a practical perspective, China has made substantial efforts to strengthen

---

[59] Deiter, C., & Rude, D. (2005). *Human security and aboriginal women in Canada.* Status of Women Canada.

[60] Hoogensen, G., Bazely, D., Christensen, J., Tanentzap, A., & Bojko, E. (2009). Human security in the Arctic-Yes, it is relevant! *Journal of Human Security, 5*(2), 1–10.

[61] Hossain, K., Martín, J. M. R., & Petretei, A. (2018). Understanding human security as a tool to promote societal security in the Arctic. In *Human and societal security in the circumpolar Arctic* (pp. 3–15). Brill.

[62] State Council Information Office of the People's Republic of China. (2018). China's Arctic Policy.

environmental security in the Arctic. For instance, China has contributed to the development of scientific research centers and the construction of modern icebreakers, both of which will enhance environmental security in the region. China has also made concerted efforts to abide by international law in the Arctic. International law stipulates rules and regulations pertaining to sovereign states, which seek to foster consistent, stable, and fair international relations. As China has pledged to obey these rules and regulations, its enterprises and people must act in accordance with international law in the Arctic, under the supervision of the Chinese government. In addition, companies have an obligation to maintain or develop sustainable practices in the Arctic region.[63] Nonetheless, China has tended to neglect several key aspects of cooperation, especially in its cooperation with Arctic countries on issues relating to human and social security, such as the preservation of local cultures and identities and the local fishing industry. This lack of awareness may create problems for China as it seeks to be more involved in the Arctic. Consequently, China should intensify its non-traditional security cooperation with other countries in the Arctic in a practical manner.

8.4 Limitations to China's Involvements in the Arctic

Despite China's financial and scientific advantages in the Arctic, it faces significant obstacles in becoming more involved in Arctic affairs. First, China has insufficient legitimacy to compete with the other two superpowers—the United States and Russia—in the Arctic, because China is a non-Arctic country with an "observer" or "outsider" status in the Arctic Council, whereas the United States and Russia are important member states. In addition, Arctic countries have expressed concerns about China's intervention in their areas.[64] Thus, China must cooperate with Arctic countries if it wishes to enhance its legitimacy in the region, such as by partnering with Russia to deliver multiple projects. In this exchange, Russia benefits from China's financial capital and its advanced

---

[63] Tillman, H., Yang, J., & Nielsson, E. T. (2018). The Polar Silk Road: China's new frontier of international cooperation. *China Quarterly of International Strategic Studies*, 4(3), 345–362.

[64] Hong, N. (2018). *China's interests in the Arctic: Opportunities and challenges* (Institute for China-American Studies White Paper).

scientific research. Simultaneously, China can adopt Russian deep-water drilling or extraction technologies, which may help boost its industrial capacities. Although Russia enjoys the economic benefits it derives from its partnership with China, it has also attempted to prevent external countries from intervening in the region and generally avoided sharing its technologies with other parties.[65]

Nonetheless, Russia has expressed concerns about China playing an official role in the Arctic Council. Moreover, bilateral relations between China and the European Union in the Arctic have become increasingly complicated. Compared with Russia, the European Union has had a longer history of cooperation with China in areas pertaining to the economy and trade.[66] In terms of political security, Russia has been a more stable and reliable partner of China than the European Union, because most member states regard China's actions in the Arctic as ambitious attempts at geopolitical expansion. State cooperation incentivized by common interest can help to generate mutually beneficial solutions for both developed and developing countries.[67,68,69] To some extent, these developments remind member states of the European Union of the Soviet Union's expansion during the twentieth century. It is difficult for China to balance its relations with the European Union and Russia, both major participants in current Arctic affairs. As the European Union and the United States often have the same opinions on specific security issues, it is difficult for the European Union and China to discuss Arctic affairs. Additionally, although China, South Korea, and Japan are geographically neighbors, because of historical problems, these countries have different interests, and their joint initiatives under the BRI have yet to achieve outstanding performance. Nonetheless, the present situation seems to be

[65] Alexeeva, O., & Lasserre, F. (2018). An analysis on Sino-Russian cooperation in the. Arctic in the BRI era. *Advances in Polar Sciences, 29*(4), 269–282.

[66] Sheng, L., and do Nascimento, D. F. (2021) *The Belt and Road initiative in South-South Cooperation: The impact on world trade and geopolitics.* Palgrave Macmillan.

[67] Sheng, L. (2010). Growth–volatility tradeoff in the face of financial openness: a perspective of developing economies. *Cambridge Review of International Affairs, 23*(4), 609–622.

[68] Sheng, L. (2011). Theorising free capital mobility: The perspective of developing countries. *Review of International Studies, 37*(5), 2519–2534.

[69] Sheng, L. (2014). Capital controls and international development: A theoretical reconsideration. *Global Policy, 5*(1), 114–120.

limited by the level of cooperation between the three East Asian countries in Arctic affairs.

Another concern about China's engagement in the Arctic is its lack of military bases in the region. Hsiung argued that the absence of military bases in the Arctic is a growing concern for Beijing.[70] China has insufficient military power to protect its infrastructure, investment projects, and maritime trade routes in the region. In addition, China's military may fail to respond in time to potential military conflicts with the United States in the Arctic. Because of this, China is completely dependent on its ally, Russia, for protection in the Arctic; however, this is not a long-term solution. Moreover, China and Russia have not maintained any military alliances in recent years. It is impractical for China to rely on Russia's military or protection, as this is inconsistent with China's independence policy. Furthermore, as a permanent member of the United Nations Security Council, China shoulders the important responsibility of jointly maintaining peace and security in the Arctic. Lacking military capacity in the Arctic, China's participation in Arctic affairs will follow the basic principles of "respect, cooperation, win–win, and sustainability" in order to achieve the goal of "understanding, protecting, using and participating in the governance of the Arctic, and safeguarding the common interests of all countries and the international community in the Arctic."[71] To a certain extent, China's claim that non-Arctic countries can participate in Arctic affairs on the basis of international law—as advocated in its Arctic policy white paper—is consistent with Russia's maintenance of its dominant position in Arctic affairs, especially with regard to the internationalization of the NRS.

Chinese companies lack the technological expertise and knowledge required for operating in the Arctic.[72] Extremely cold weather and numerous icebergs represent new challenges for Chinese companies, which have never undertaken deep-water operations. These companies thus need to invest heavily in advanced machinery and technologies for surveying and extracting natural minerals. Such high start-up costs have

---

[70] Weidacher Hsiung, C. (2016). China and Arctic energy: Drivers and limitations. *The Polar Journal, 6*(2), 243–258.

[71] Sheng, L., & Zhang, Z. Y. (in press). Revisiting global imbalances: A comparative analysis of income and consumption inequality. Economics & Politics.

[72] Weidacher Hsiung, C. (2016). China and Arctic energy: Drivers and limitations. *The Polar Journal, 6*(2), 243–258.

prevented them from participating in resource extraction in the Arctic.[73] In addition, an analysis of China and Russia's energy systems revealed that "the Chinese industry has a very high potential for development with low-efficiency results."[74] Hsiung argued that the Yamal LPG project, which is jointly operated by Russia and China, involves high risks and high costs and that its profitability cannot be guaranteed unless it receives full support from the Russian government.[75] Therefore, the extent to which Chinese enterprises and the Chinese government will reap economic benefits from remains unclear.

Finally, China's involvement in the Arctic is likely to be limited by the fact that the majority of the Arctic countries are allies of the United States. Furthermore, Sino–American relations have deteriorated in recent years and have not improved under the Biden administration. On his first day in office, President Biden set the tone for the United States' environmental policy by rejoining the Paris Agreement and imposing a temporary moratorium on oil leasing in the Arctic National Wildlife Refuge of Alaska. Nevertheless, the Biden administration will pursue other policies that were introduced under the Trump administration, such as the enhancement of the country's military and civilian capabilities to operate in the harsh Arctic climate.[76] Additionally, during a visit to Iceland and Greenland in May 2021, Biden's top diplomat noted that the military would expand its presence and increase the frequency of its joint military exercises with its NATO allies in the "high north."[77] This series of aggressive, confrontational and escalating actions and signals by the United States indicates that the future of international relations in the Arctic region is likely to be complicated and challenging. Hence, it is uncertain whether there will be future collaborations between China and the Arctic countries and whether the Arctic countries will accept China as a partner in the region.

---

[73] Ibid.

[74] Steblyanskaya, A., Qingchao, X., Razmanova, S., Steblyanskiy, N., & Denisov, A. (2021). China and Russia energy strategy development: Arctic LNG. *International Journal of Energy Economics and Policy, 11*(4), 450–460.

[75] Weidacher Hsiung, C. (2016). China and Arctic energy: Drivers and limitations. *The Polar Journal, 6*(2), 243–258.

[76] Carnegie Endowment for International Peace. (2021, May 17). *A fresh start on U.S. Arctic Policy under Biden.* Carnegie Moscow Center.

[77] Foreign Policy. (2021, May 20). Biden positions U.S. as player in the Arctic.

References

Alexeeva, O., & Lasserre, F. (2018). An analysis on Sino-Russian cooperation in the. Arctic in the BRI era. *Advances in Polar Sciences, 29*(4), 269–282.

Arctic Council. (1996). Declaration on establishment of the Arctic Council (The Ottawa Declaration) 1996. http://library.arcticportal.org/1541/1/00_ottawa_decl_1996_signed.pdf

Arctic Portal. (2013). *China–Nordic Arctic Research Center.* https://arctic portal.org/ap-library/news/1136-china-nordic-arctic-research-center-inaugu rated

Bennett, M. M. (2015). How China sees the Arctic: Reading between extrare-gional and interregional narratives. *Geopolitics, 20*(3), 645–668.

Biedermann, R. (2020). The Polar Silk Road: China's multilevel Arctic strategy to globalize the Far North. *Contemporary Chinese Political Economy and Strategic Relations, 6*(2), 571–615.

Blaxekjær, L. Ø., Lanteigne, M., & Shi, M. (2018). The Polar Silk Road & the West Nordic Region. *Arctic Yearbook, 2018*, 437–455.

Blunden, M. (2012). Geopolitics and the northern sea route. *International Affairs, 88*(1), 115–129.

Borgerson, S. (2008). Arctic meltdown: The economic and security implications of global warming. *Foreign Affairs, 87*(2), 63–77.

Campbell, C. (2013). *China and the Arctic: Objectives and obstacles.* US-China. Economic and Security Review Commission Research Report. http://library.arcticportal.org/1677/1/China-and-the-Arctic_Apr2012.pdf

Carnegie Endowment for International Peace. (2021, May 17). *A fresh start on U.S. Arctic policy under Biden.* Carnegie Moscow Center. https://carnegiem oscow.org/commentary/84543

Chinadaily. (2018). *Finland touts "Arctic Corridor" tunnel.* http://www.chinad aily.com.cn/a/201803/02/WS5a984a41a3106e7dcc13efe8.html

Deiter, C., & Rude, D. (2005). *Human security and aboriginal women in Canada.* Status of Women Canada Policy Research. https://publications.gc.ca/site/eng/285465/publication.html

Dobson, A. (2007). *Green political thought.* Routledge.

Embassy of the People's Republic of China in the Republic of Iceland. (2012, April 25). *Chinese premier Wen Jiabao pays official visit to Iceland.* http://is.china-embassy.org/eng/xwdt/t926273.htm

Foreign Policy. (2021). *Biden positions U.S. as player in the Arctic.* Retrieved September 21, 2021, from https://foreignpolicy.com/2021/05/20/biden-arctic-geopolitics-russia-china/

Greaves, W. (2016). *Thinking critically about security and the Arctic in the Anthropocene.* The Arctic Institute. https://www.thearcticinstitute.org/thi nking-critically-security-arctic-anthropocene/

Havnes, H. (2020). The Polar Silk Road and China's role in Arctic governance. *Journal of Infrastructure, Policy and Development, 4*(1), 121–138.

Ho, J. (2010). The implications of Arctic sea ice decline on shipping. *Marine Policy, 34*(3), 713–715.

Hong, N. (2018). *China's interests in the Arctic: Opportunities and challenges* (Institute for China-American Studies White Paper). http://chinaus-icas.org/wp-content/uploads/2018/03/2018.03.06-China-Arctic-Report.pdf

Hoogensen, G., Bazely, D., Christensen, J., Tanentzap, A., & Bojko, E. (2009). Human security in the Arctic-yes, it is relevant! *Journal of Human Security, 5*(2), 1–10.

Hossain, K., Martín, J. M. R., & Petretei, A. (2018). Understanding human security as a tool to promote societal security in the Arctic. In *Human and societal security in the circumpolar Arctic* (pp. 3–15). Brill.

Hossain, K., Zojer, G., Greaves, W., Roncero, J., & Sheehan, M. (2017). Constructing Arctic security: An inter-disciplinary approach to understanding security in the Barents region. *Polar Record, 53*(1), 52–66.

Jiang, Y. A. (2019). Multilateral cooperation on the Polar Silk Road: Opportunities, challenges and development paths. *Pacific Journal* (Taipingyang Xuebao), *27*(8), 67–77. https://doi.org/10.14015/j.cnki.1004-8049.2019.08.006

Kopra, S. (2020). China, great power responsibility and Arctic security. In *Climate change and Arctic security* (pp. 33–52). Springer.

Kossa, M. (2019). China's Arctic engagement: Domestic actors and foreign policy. *Global Change, Peace and Security, 32*(1), 19–38.

Li, X. P., Fang, Z., & Liu, H. Y. (2017). Rights and obligations of the Arctic Region under the United Nations Convention on the law of the sea and their relationship with geographically disadvantaged States. *Polar Research* (2), 279–285.

Li, Z. F., & Peng, Y. (2018). "The Polar Silk Road" and the Great Arctic Network: Role, evolution and China's strategy. *Northeast Asian Economic Research, 2*(5), 5–20.

Lim, K. S. (2018). China's Arctic policy and the Polar Silk Road vision. *Arctic Yearbook 2018*, 420–432.

Liu, N. (2018). Will China build a green Belt and Road in the Arctic? *Review of European, Comparative & International Environmental Law, 27*(1), 55–62.

Ministry of Ecology and Environment of the People's Republic of China. (2017). *Guidance on promoting Green Belt and Road.* http://english.mee.gov.cn/Resources/Policies/policies/Frameworkp1/201706/t20170628_416864.shtml

Ministry of Economic Development of the Russian Federation. (2021). Russia boosts Arctic LNG shipments to China to meet growing energy demand. Russia boosts Arctic LNG shipments to China to meet growing energy demand. http://www.ved.gov.ru/eng/general/news/19/28775.html

Mohr, J. (2020). China in the Arctic and the case of Greenland. In *Handbook on geopolitics and security in the Arctic* (pp. 113–129). Springer.

Pelaudeix, C. (2018). *Along the road: China in the Arctic*. EUISS Briefs. https://www.iss.europa.eu/content/along-road-%E2%80%93-china-arctic

Pincus, R. (2020). Three-way power dynamics in the Arctic. *Strategic Studies Quarterly, 14*(1), 40–63.

Reinke de Buitrago, S. (2020). China's aspirations as a "Near Arctic State": Growing stakeholder or growing risk? In *Handbook on geopolitics and security in the Arctic* (pp. 97–112). Springer.

Sheng, L. (2010). Growth–volatility tradeoff in the face of financial openness: A perspective of developing economies. *Cambridge Review of International Affairs, 23*(4), 609–622.

Sheng, L. (2011). Theorising free capital mobility: The perspective of developing countries. *Review of International Studies, 37*(5), 2519–2534.

Sheng, L. (2014a). Capital controls and international development: A theoretical reconsideration. *Global Policy, 5*(1), 114–120.

Sheng, L. (2014b). Economic structure, cost outsourcing and global imbalances. *Journal of Australian Political Economy, 74*, 81–95.

Sheng, L. (2014c). Effects of foreign expansion on local growth: The case of Macao. *European Planning Studies, 22*(8), 1735–1743.

Sheng, L. (2015). Theorizing income inequality in the face of financial globalization. *The Social Science Journal, 52*(3), 415–424.

Sheng, L. (2021). *How Covid-19 reshapes new world order: Political economy perspective*. Springer.

Sheng, L., & do Nascimento, D. F. (2021a). *The Belt and Road Initiative in South-South cooperation: The impact on world trade and geopolitics*. Palgrave Macmillan.

Sheng, L., & do Nascimento, D. F. (2021b). *Love and trade war: China and the US in historical context*. Springer.

Sheng, L., & Share, M. (2022). Exchange between Russia and the Guangdong-Hong Kong–Macao Greater Bay Area in a historical context. *Revista de Cultura, 67*, 46–59.

Sheng, L., & Zhang, Z. Y. (in press). Revisiting global imbalances: A comparative analysis of income and consumption inequality. *Economics & Politics*.

State Council Information Office of the People's Republic of China. (2018). *China's Arctic Policy*. http://english.www.gov.cn/archive/white_paper/2018/01/26/content_281476026660336.htm

Steblyanskaya, A., Qingchao, X., Razmanova, S., Steblyanskiy, N., & Denisov, A. (2021). China and Russia energy strategy development: Arctic LNG. *International Journal of Energy Economics and Policy, 11*(4), 450–460.

Tillman, H., Yang, J., & Nielsson, E. T. (2018). The Polar Silk Road: China's new frontier of international cooperation. *China Quarterly of International Strategic Studies, 4*(3), 345–362.

Wang, H. J., Chen, H. P., & Liu, J. P. (2015). Arctic sea ice decline intensified haze pollution in eastern China. *Atmospheric and Oceanic Science Letters, 8*(1), 1–9.

Weber, J. (2020). Limited cooperation or upcoming alliance? Russia, China and the Arctic. In *Handbook on geopolitics and security in the Arctic* (pp. 345–361). Springer.

Weidacher Hsiung, C. (2016). China and Arctic energy: Drivers and limitations. *The Polar Journal, 6*(2), 243–258.

Woon, C. Y. (2020). Framing the "Polar Silk Road": Critical geopolitics, Chinese scholars and the (re)positionings of China's arctic interests. *Political Geography, 78.* https://doi.org/10.1016/j.polgeo.2019.102141

XinHuaNet. (2019, November 25). *China's polar icebreaker Xuelong 2 finished its icebreaking tasks.* Retrieved September 26, 2021, from http://www.xinhua net.com/english/2019-11/25/c_138581294.htm

XinHuaNet. (2021, May 7). *Chinese research icebreaker Xuelong 2 completes Antarctic expedition.* Retrieved September 27, 2021, from http://www.xin huanet.com/english/2021-05/07/c_139930634.htm

Yang, J. (2018) The international environment and response to the construction of the Polar Silk Road. *People's Forum· Academic Frontier* (Renmin Luntan· Xueshu Qianyan), (11), 13–23. https://doi.org/10.16619/j.cnki.rmltxsqy. 2018.11.002

Yearender: Growing China brings prosperity to Eurasia. Xinhua. (2017, December 31). Retrieved September 14, 2021, from http://www.xinhuanet. com/english/2017-12/31/c_136863413.htm

Yilmaz, S. (2017). Exploring china's Arctic strategy: Opportunities and challenges. *China Quarterly of International Strategic Studies, 3*(1), 57–78.

Yilmaz, S., & Liu, C. M. (2019). Remaking Eurasia: The Belt and Road Initiative and China-Russia strategic partnership. *Asia Europe Journal, 18*(3), 259–280.

Yin, Y., & Sheng, L. (2021). Theorizing about global imbalances: An inequality perspective. *Argumenta Oeconomica, 46,* 169–181.

Zheng, Y. Q. (2019). China-Nordic blue economic passage: Basis, challenges and paths. *China International Studies,* (5), 29–49.

Zhu, X. P., Zhang, Y. F., Liu, X., & Wang, Y. G. (2018). Implementing the spirit of the 19th National Congress to build the "Ice Silk Road"-Jilin University-Russian Academy of Military Sciences "Ice Silk Road" Seminar. *Northeast Asia Forum, 27*(2), 3–7.

Zou, Y., Wang, Y., Zhang, Y., & Koo, J. H. (2017). Arctic sea ice, Eurasia snow, and extreme winter haze in China. *Science Advances, 3*(3), e1602751. https://doi.org/10.1126/sciadv.1602751

CHAPTER 9

Conclusion

Abstract The main aim of this book is to elaborate on the participation and interaction of the three great powers—i.e., China, Russia, and the United States—in Arctic affairs from an international relations perspective. The rapid warming of the Arctic has ushered in a new era of development in the region. Although there are still some divergences, China and Russia are expanding and deepening their cooperation in the Arctic region. There are broad prospects for further Arctic cooperation as it undergoes a dynamic adjustment.

Keywords Arctic governance · Climate change · Great powers

The Arctic has long been an arena for games among the world's biggest powers. Throughout the past century, the United States and Russia have dominated regional affairs in the Arctic. Over the past decades, the two powers' policies and strategies toward the Arctic have continuously been adjusted to meet the economic and geopolitical demands of different periods. The geopolitical value of the Arctic has increased rapidly in recent years owing to the melting of the polar ice caps with global warming. More than ever, the Arctic is attracting the attention of the world's major powers, and competition in the region is intensifying.

E. L. Sheng, *Arctic Opportunities and Challenges*, https://doi.org/10.1007/978-981-19-1246-7_9

In the post-Cold War era, the Arctic enjoys a high degree of geopolitical stability, with neither military conflicts nor fierce boundary disputes in the region. However, Arctic affairs remain complicated and intricate because they often involve different factors, such as environmental changes, geopolitical games, security, and international law. Current Arctic affairs often reflect the integration of traditional security and non-traditional security, as well as the contradiction between national governance and global governance. That is, Arctic affairs have become more complex and unpredictable. Climate change has cemented the strategic position of the Arctic in the plans of various countries for resource development and their security strategies. With global warming and geopolitical rivalries, the situation in the Arctic is likely to become increasingly complex as more and more internal and external players enter the game.

At the same time, global warming not only offers new opportunities for development to the immediate Arctic region and local populations but also introduces many new and unknown challenges and triggers profound changes in international politics. One after another, the various Arctic states have issued new policies of engagement with the Arctic. Notably, these countries focus their policies on several common foundations, such as interest in resources, environmental protection, and regional economic and social development objectives. Of the players in the Arctic, the United States, Russia, and China display the most extensive security interest. While seeking to strengthen their means of competition, these countries also emphasize the principle of international governance. However, the scope of such governance is usually limited to other Arctic countries. This demonstrates the limited and exclusive nature of international cooperation in the Arctic.

According to their geographic proximity to the Arctic, we argue that there are presently four kinds of players in the Arctic game: core players, direct players, semi-direct players, and extra-regional players. First, as the two largest powers that have dominated the region for over half a century, the United States and Russia are core players in the game. Second, the other Arctic Council members—Canada, Demark, Norway, Sweden, Finland, and Iceland—are direct players owing to their membership in the Arctic Council and their respective claims over territories in the Arctic. Third, the European Union, which has membership in the Arctic Council, is a semi-direct player that can influence Arctic Council members either directly or indirectly. Finally, extra-regional players such as China, Japan, and South Korea have played an important role in regional Arctic affairs in recent years, although these nations have no claims over the Arctic region. Despite their geographical disadvantages, the roles of China, Japan, Singapore, and South Korea in developing the Arctic cannot be overlooked because of their importance in the world economy and global trade. Especially after China joined the World Trade Organization, the country has witnessed remarkable rates of output growth and export in the globalization context.[1] China has been a growing influence on other developing economies through economic reforms, trade liberalization, and soft power presence.[2]

The rapid warming of the Arctic has also ushered in a new era of development in the region. This will have extensive strategic impacts on Arctic countries and other regions of the world. For instance, Arctic countries have formulated long-term strategies and implemented countermeasures in preparation for development opportunities triggered by changes in the Arctic's natural environment. The changing natural environment has also led to profound changes in international politics and complicated international relations. The development of the Arctic not only demands cooperation but also competition. That is, Arctic countries should be unwilling to fall behind, aim to strengthen their capacities for competition, and compete for dominance by setting the Arctic policy. The model of competition among Arctic countries has changed from one of fragmentary collision and friction to more comprehensive and systematic competition; geopolitical competition in this region has also entered a new phase of development. Although it is not an Arctic state, China is reluctant to be excluded from the intense competition for territory and resources in the Arctic.

As a typical representative of foreign countries in the region, China's Arctic strategy is circuitous and fails to conceal its strategic thirst for Arctic resources. On the one hand, China is trying to shape the Arctic into global commons and to make Arctic governance an important part of global governance. If it succeeds, then all responsible actors in the international arena will be able to participate in Arctic governance. Such governance will cover a wide range of issues, including traditional and

---

[1] Sheng, L. (2014b). Economic structure, cost outsourcing and global imbalances. *Journal of Australian Political Economy, 74*, 81–95.

[2] Sheng, L. (2014a). Capital controls and international development: A theoretical reconsideration. *Global Policy, 5*(1), 114–120.

non-traditional security affairs, the protection and use of public resources in the Arctic, and the pursuit of sustainable practices in areas such as resource exploitation, waterway development, scientific investigation, and many others. On the other hand, China is also trying to shape its image as a near-Arctic state and to constantly increase its exposure to and influence in Arctic affairs.

Driven by a domestic energy shortage and external threats to its traditional maritime trade routes, China has eagerly explored ways to diversify its suppliers and transportation routes to meet its energy needs. Facing similar national challenges, three other East Asian countries— Japan, Singapore, and South Korea—have also expressed their willingness to participate in regional affairs to maximize their national interest. In fact, China, Japan, and South Korea have participated in a variety of Arctic affairs spanning the Arctic Council, scientific research, and others.

China considers its involvement in Arctic affairs particularly important, given its intensifying competition with the United States and other Western countries. China's BRI firmly states that the country will continue to enhance its energy supply by all means[3]; this includes China's strategy to develop the Arctic. However, as an extra-regional player with limited power, China has little choice but to strengthen its cooperation with Arctic countries. China has expressed its willingness to enhance its cooperation with member states of the Arctic Council and actively called for multilateral and global governance in the North Pole. However, given that China's relations with North European members of the Council are deeply influenced by the United States, Russia represents an ideal partner for China's cooperative efforts in the Arctic. The reasons for this are as follows.

First, both China and Russia face the same geopolitical pressures from the United States and its Western allies, and the two neighboring powers have extensive common ground, upon which they can deepen their bilateral ties on security issues such as survival space and the supply of energy. In this context, the relations between China and Russia as mutual strategic partners have remained relatively stable in recent years. Second, domestic politics in both countries have remained stable under strong leadership and facilitated sustainability and policy consistency. To a large extent, the

---

[3] Sheng, L., & do Nascimento, D. F. (2021a). *Love and trade war: China and the US in historical context*. Palgrave MacMillan.

two factors discussed above lay a solid foundation for bilateral cooperation between China and Russia in the North Pole. At the same time, both China and Russia have a desire to strengthen their connections with Eurasia; this was also the reason why China implemented the BRI while Russia established the Eurasia Economic Union. In this context, the Polar Silk Road represents a bridge that serves to combine the ambitions of two giants. The proposed Polar Silk Road is an important moment in bilateral cooperation between China and Russia on Arctic affairs. In establishing this common ground, financial investments and infrastructure projects in the Arctic are expected to provide greater integration and complementarity between the two neighbors. With the continuous deepening of Sino–Russian cooperation, China will have good opportunities to become more involved in Arctic affairs.

However, the promising relations between China and Russia also involve uncertainties and even risks, which may hinder cooperation between the two countries. Such problems cannot be neglected. First, the "yoke of history" such as lost territory and the Sino–Soviet split continues to haunt Sino–Russian relations. At the same time, the two countries have diverging views on Arctic development, which can undermine mutual trust between the two sides. While Russia seeks unitary dominance in the region, China calls for multilateral governance. China cannot turn a blind eye to Russia's suspicion of its increasing presence in the region, which has led to some unpleasant outcomes in cooperation—for instance, China has received uneven divisions of interest and inactive technical support from Russia. Aside from the above factors, tensions between Russia's domestic politics and interest groups have introduced greater uncertainty to Sino–Russian Arctic development.

In the meantime, China should be cautious of "the spillover effect" from its cooperation with Russia. On the one hand, in the current context of deteriorating relations between China and the United States, the deepening of ties between Beijing and Moscow could trigger the concern of the White House.[4] After all, the United States is one of the direct shareholders in the region. On the other hand, Russia presently unilaterally controls the Northeast waterway and manages disputes over the territory and jurisdiction of maritime routes in the Arctic Ocean. These factors dictate the relations between the member states of the Arctic

[4] Sheng, L. (2021c). *How Covid-19 reshapes new world order: Political economy perspective.* Springer.

Council. Cooperation between China and Russia in developing the Arctic waterway may worry other countries in the region and subsequently influence China's relations with Northern European countries.[5] In this case, China's Arctic ambitions will be further emphasized in the current complex and tense international situations.

It is very important that China seeks a balance in its Arctic affairs. This balance should exist between China and the Arctic states, especially Russia and the United States, and be reflected in China's strategic arrangement for the Arctic. Specifically, in expanding its influence in Arctic affairs, China needs to constantly cooperate with Arctic countries, especially when the United States sees China as its main competitor and Russia has doubts about all-round cooperation with China. China's strategy for dealing with Arctic affairs is equally critical. As an observer on the Arctic Council, China's ability to set an agenda depends more on its leadership, which is characterized by comprehensive strength and the trust of Arctic states. China is motivated to diversify global shipping and continuously enhance its influence by promoting the Polar Silk Road. However, this initiative has also raised concerns in many countries. China must therefore strike a delicate balance as it seeks to improve its capacity for policy intervention while reducing its external resistance.

In conclusion, as an outsider to the Arctic region with extensive geographical disadvantages, China adopts a long-term strategy in its involvement in Arctic affairs. China is likely to face an increasingly intricate and complex situation in Arctic affairs involving both competition and challenges. However, given the currently intensifying competition between China and the United States, it is necessary for China to coordinate its relations (regarding Arctic affairs) with the United States and its Western allies. Although China and Russia have diverging views and have experienced friction in their bilateral relations pertaining to the development of the Arctic, their relations continue to undergo dynamic adjustments and show promise in facilitating prosperous Sino–Russian relations and the successful development of the BRI.

---

[5] Sheng, L., & do Nascimento, D. F. (2021b). *The belt and road initiative in South–South cooperation: The impact on world trade and geopolitics*. Palgrave Macmillan.

References

Sheng, L. (2014a). Capital controls and international development: A theoretical reconsideration. *Global Policy, 5*(1), 114–120.

Sheng, L. (2014b). Economic structure, cost outsourcing and global imbalances. *Journal of Australian Political Economy, 74*, 81–95.

Sheng, L. (2021c). *How Covid-19 reshapes new world order: Political economy perspective*. Springer.

Sheng, L., & do Nascimento, D. F. (2021a). *Love and trade war: China and the US in historical context*. Palgrave Macmillan.

Sheng, L., & do Nascimento, D. F. (2021b). *The belt and road initiative in South–South cooperation: The impact on world trade and geopolitics*. Palgrave Macmillan.

INDEX